Giuseppe Lippi • Gianfranco Cervellin • Emmanuel J. Fa

In Vitro a..

Patient Safety

Edited by
Oswald Sonntag and Mario Plebani

Volume 4

Giuseppe Lippi
Gianfranco Cervellin
Emmanuel J. Favaloro
Mario Plebani

In Vitro and In Vivo Hemolysis

An Unresolved Dispute
in Laboratory Medicine

DE GRUYTER

Authors

Giuseppe Lippi, MD
U.O. Diagnostica Ematochimica
Dipartimento di Patologia e Medicina di Laboratorio
Azienda Ospedaliero-Universitaria di Parma, Italy
Email: glippi@ao.pr.it

Gianfranco Cervellin, MD
U.O. di Pronto Soccorso e Medicina d' Urgenza
Azienda Ospedaliero-Universitaria di Parma, Italy
Email: gcervellin@ao.pr.it

Emmanuel J. Favaloro, PhD
Department of Haematology
Institute of Clinical Pathology and Medical Research (ICPMR)
Westmead Hospital, Australia
Email: emmanuel.favaloro@swahs.health.nsw.gov.au

Mario Plebani, MD
Dipartimento di Medicina di Laboratorio
Università di Padova, Italy
Email: mario.plebani@unipd.it

ISBN 978-3-11-024613-1 • e-ISBN 978-3-11-024614-8

Library of Congress Cataloging-in-Publication Data
A CIP catalog record for this book has been applied for at the Library of Congress

Bibliographic information published by the Deutsche Nationalbibliothek
The Deutsche Nationalbibliothek lists this publication in the Deutsche Nationalbibliografie; detailed bibliographic data are available in the Internet at http://dnb.dnb.de.

Typesetting: Apex CoVantage, LLC
Printing: Hubert & Co. GmbH & Co. KG, Göttingen
Cover image: Comstock/Getty Images.

♾ Printed on acid-free paper
Printed in Germany
www.degruyter.com

Contents

Preface

Hemolysis is an important phenomenon in laboratory medicine because it may derive from two different sources, which deserve different approaches: (a) *in vivo* hemolysis, which may be caused by a variety of conditions and disorders, can lead to various degrees of anemia (up to life-threatening anemia, when the concentration of hemoglobin declines very rapidly and/or falls below 0.06 g/L), and (b) *in vitro* hemolysis, which is caused by inappropriate procedures for collection and/or handling of the biological specimen, and can seriously impact patient care and a laboratory's reputation through a wide range of adverse affects on test results. Hemolytic specimens are a frequent event in laboratory practice, with an average prevalence described as about 8.8% of all of routine samples referred to a clinical laboratory, and accounting also for 39–69% of all the unsuitable specimens received in clinical laboratories (i.e., nearly five times higher than the second major cause). Hemolysis, therefore, still represents the most prevalent preanalytical error across countries, health care facilities, and types of clinical laboratories. Several causes have been traditionally associated with an increased burden of *in vitro* hemolysis, including difficult venipunctures, use of inappropriate devices, underfilling of blood tubes, exposure to extreme temperatures and physical forces during sample transportation via pneumatic systems, and centrifugation at a too high speed of partially coagulated specimens. In addition, even an excessive shaking or mixing of blood after collection (i.e., for times longer than recommended or with great force) is also usually acknowledged as a leading source of RBC (red blood cell) injury. The prevalence of hemolytic specimens is increasingly considered a reliable index for assessing preanalytical quality and, more interestingly, for introducing new tools and guidelines for the safe management of hemolytic samples.

Therefore, this volume represents a fundamental source for updating knowledge on hemolysis and on a valuable approach to hemolytic samples in clinical practice.

The authors
March 2012

1 Structure and function of red blood cells

Red blood cells (RBCs, also referred to as **erythrocytes**), are the most common type of blood cell and the vertebrate organism's principal means of delivering oxygen to the body's tissues via the flow of blood through the circulatory system. The main function of **RBCs** is to capture oxygen in the lungs and, as the RBCs squeeze through even the tiniest blood vessels, to release it as required to the peripheral tissues. The main component of an RBC is **hemoglobin**, an iron-containing biomolecule that can bind oxygen and is responsible for the blood's red color. In humans, mature RBCs are flexible biconcave disks that lack a cell nucleus and most organelles. These cells originate from the bone marrow and are characterized by a life span of some 90 to 120 days before they undergo degradation by a process of phagocytosis in the reticuloendothelial system of the spleen, liver, and bone marrow. The processes of production and degradation typically occur at the same rate, thereby providing a balance in the total number of circulating RBCs or consistency in the RBC count [1,2].

The process leading to the development of RBCs is termed **erythropoiesis** and lasts nearly 7 days. By it, erythrocytes are continuously produced in the red bone marrow of large bones at the staggering rate of about 2 million per second in a healthy adult. The process is regulated in two ways by the hormone erythropoietin (EPO), which can act both to stimulate production of the cells from their cell precursors (stem cells) in the bone marrow and to prevent apoptosis of immature RBCs (**reticulocytes**) by a process commonly referred to as neocytolysis. In adults, the reticulocytes make up nearly 1% of circulating RBCs [1,2].

The size of RBCs varies widely among vertebrate species. The erythrocyte's width is, on average, about 25% greater than the diameter of a capillary, and it has been hypothesized that this feature improves the transfer of oxygen from erythrocytes to tissues. A typical human RBC has a diameter of 6 to 8 μm and is about 2 μm thick. The cells consequently have a volume of about 90 fL with a surface area of about 136 μm² and can swell up to a spherical shape containing 150 fL without membrane distention (▶Fig. 1.1).

Mammalian erythrocytes lack a cell nucleus and are typically shaped in the form of a biconcave disk, being flattened and depressed in the center, with a dumbbell-shaped cross section and a torus-shaped rim on the edge of the disk. The only vertebrates without RBCs are the crocodile icefishes (family *Channichthyidae*), which live in very oxygen-rich cold water and thereby transport oxygen freely dissolved within their blood. The original biconcave shape of erythrocytes is thought to optimize the rheological properties of blood in the large vessels, such as maximization of laminar flow and minimization of platelet scatter (▶Fig. 1.1). In adult life, humans have nearly 2 to 3×10^{13} RBCs, constituting approximately one quarter of the total number of human body cells. The number of erythrocytes is therefore much higher than that of other blood cells (from 4.2 to 6.2×10^{12}/L in males and from 3.8 to 5.5×10^{12}/L in females, versus 4.0 to 11.0×10^{9}/L white blood cells and 150,000 to 400,000 10^{9}/L platelets).

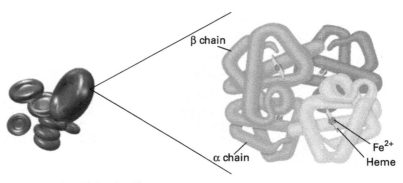

Fig. 1.1: Structure of red blood cells (RBCs).

The production of RBCs from the bone marrow is finely regulated by the hormone Epo, a 165–amino acid glycoprotein with a molecular weight of 34 kDa containing approximately 40% carbohydrates. The hormone is mainly produced by the peritubular capillary endothelial cells of the kidney and, to a lesser extent, of the liver, as well as by macrophages of the bone marrow. The relative contribution of these extrarenal sites to the total production of erythropoietin in adults with severe hypoxia ranges from 10% to 15%. Hypoxia represents the main trigger of erythropoietin production and stimulates both synthesis and secretion of the hormone from the kidney through a complex pathway involving sensors that are highly sensitive to the blood's oxygen tension. At variance with other hematopoietic growth factors, erythropoietin selectively targets specific membrane receptors of erythroid progenitor cells, especially the burst forming unit–erythroid (BFU-E) and the colony forming unit–erythroid (CFU-E) cells, whereas sensitivity to the hormone becomes lost in the proerythroblast stage. The erythropoietin-mediated signaling in both CFU-E and BFU-E cells stimulates the formation of erythroblast colonies within 7 and 15 days, respectively. The second leading mechanism by which erythropoietin modulates erythropoiesis is by counteracting the selective apoptosis of young circulating RBCs by neocytolysis. Basically erythropoietin allows the survival of progenitor cells, especially CFU-E cells. Therefore the increased production of erythropoietin in patients with anemia or other causes of tissue hypoxia prevents the erythropoietin-dependent programmed cell death of erythroid progenitors. This is considered a compensatory mechanism that allows the survival of the appropriate number of RBCs [1,2].

As mentioned, the main constituent of RBCs is hemoglobin, a complex metalloprotein containing heme groups, whose iron atoms temporarily bind to oxygen molecules (O_2) (▶Fig. 1.1). Each human erythrocyte contains nearly 270 million hemoglobin molecules, each carrying four heme groups, so that hemoglobin represents approximately one third of the total RBC volume. It is in the capillaries of peripheral tissues that O_2 diffuses through the RBC membrane. Hemoglobin in RBCs also conveys carbon dioxide (i.e., the end product of oxygen metabolism in peripheral tissues), which is transported back to the pulmonary capillaries of the lungs to then be expelled by breathing. Isolated blood plasma is typically straw-colored and the hemoglobin confers a red color to blood, RBCs themselves can change color depending on the state of the hemoglobin. Thus, when combined with oxygen, the resulting oxyhemoglobin is scarlet, whereas when oxygen has been released, the resulting deoxyhemoglobin is of a dark red burgundy color, appearing bluish through the vessel wall and skin [1,2].

2 Red blood cell parameters

In clinical and laboratory hematology, the structure and metabolism of RBCs are described through a variety of parameters, which basically include the RBC count, hemoglobin concentration, **hematocrit**, mean corpuscular volume (MCV), RBC distribution width (RDW), mean corpuscular hemoglobin (MCH), and mean corpuscular hemoglobin concentration (MCHC). Typical reference ranges for these parameters are shown in ▶Tab. 2.1.

2.1 Hematocrit

The term **hematocrit** (Hct or HCT) – or packed cell volume (PCV) or erythrocyte volume fraction (EVF) – was coined in 1903 from the Greek words *hema* (blood) and *krites* (which means gauging or judging the blood). The hematocrit basically represents the proportion (or fraction) of the blood volume that is occupied by RBCs. The hematocrit is normally between 0.41 and 0.53 in males and between 0.36 and 0.46 in females. With modern laboratory instrumentation, the hematocrit is not directly measured; rather, it is calculated by multiplying the RBC count by the MCV. The figure arrived at in this way, rather than by direct measurement of the PCV, is slightly more accurate, because the direct method includes small amounts of blood plasma trapped between the red cells.

2.2 Mean corpuscular volume

The mean corpuscular volume (MCV), also known as mean cell volume, is a measure of the size of the average RBC volume and is reported as part of a standard complete blood count. In patients with anemia, the MCV measurement allows classification of anemias into normocytic (MCV within normal range), microcytic (MCV below normal range) or macrocytic (MCV above normal range). This parameter can be directly measured (e.g., in volume-sensitive automated blood cell counters) or be estimated by a calculation from an independently measured RBC count and hematocrit (i.e., by dividing the hematocrit by the RBC count).

2.3 Mean corpuscular hemoglobin

The mean corpuscular hemoglobin (MCH), or "mean cell hemoglobin," is the average mass of hemoglobin per RBC. This parameter is calculated by dividing the total mass of hemoglobin by the number of RBCs in a given volume of blood (i.e., MCH = (hemoglobin*10)/RBC). The normal value in humans is 27 to 31 picograms (pg) per cell.

Tab. 2.1: Reference ranges for red blood cell (RBC) parameters

Test	Gender	Lower limit	Upper limit	Unit
Hemoglobin (Hb)	Male	8.4	10.9	mmol/L
		135	175	g/L
	Female	7.5	9.9	mmol/L
		120	160	g/L
Free plasma hemoglobin		0.16	0.62	µmol/L
		1.0	4.0	mg/dL
Hematocrit (Hct)	Male	0.41	0.53	
	Female	0.36	0.46	
	Child	0.31	0.43	
Mean cell volume (MCV)	Male	82	102	fL
	Female	78	101	fL
RBC distribution width (RDW)		11.5	14.5	%
Mean cell hemoglobin (MCH)		1.53	2.16	fmol per cell
		25	35	pg per cell
Mean corpuscular hemoglobin concentration (MCHC)		31	35	g/dL
		19.3	21.7	mmol/L
Erythrocytes/red blood cells (RBCs)	Male	4.3	5.9	$\times 10^{12}$/L
	Female	3.5	5.5	$\times 10^{12}$/L
	Infant/ child	3.8	5.5	$\times 10^{12}$/L
Reticulocytes		26	130	$\times 10^{9}$/L
	Adult	0.5	1.5	% of RBCs
	Newborn	1.1	4.5	% of RBCs
	Infant	0.5	3.1	% of RBCs

2.4 Mean corpuscular hemoglobin concentration

The mean corpuscular hemoglobin concentration (MCHC) is a measure of the concentration of hemoglobin in a given volume of packed RBCs, which is calculated by dividing the hemoglobin by the hematocrit.

2.5 Red blood cell distribution width

The RBC distribution width (RDW) is a measure of the variation of RBC width. The RDW is usually calculated using the following mathematical formula: RDW = (standard deviation of MCV / mean MCV) × 100. Whereas the normal reference range in human RBCs is 11% to 14% and although RBCs usually are of a standard size (i.e., from 6 to 8 µm), certain disorders can cause a significant variation in cell size. Thus a higher RDW value simply indicates greater variation in cell size. Therefore a folate or vitamin B12 deficiency typically produces a macrocytic anemia with a normal RDW, whereas iron deficiency anemia initially presents with a varied size distribution of RBCs and therefore an increased RDW. Finally, an elevated RDW, representing a range of RBCs of unequal sizes, is known as anisocytosis. It is important to mention here, however, that the term *width* in RDW can actually be considered misleading because it does not refer directly to the diameter of the RBCs but is instead a measure of the deviation of their volume.

3 Definition of hemolysis

The word **hemolysis** (also known as **haemolysis**), derives from the Greek *hemo* (i.e., blood), and *lysis* (i.e., breakdown). Hemolysis is a pathological process characterized by the breakdown of RBCs, with the resulting release of hemoglobin and other intracellular components into the surrounding fluid [3,4]. Hemolysis is an important phenomenon in medicine for many reasons, the two main ones being (a) that in vivo hemolysis, which may be caused by a variety of conditions and disorders, can lead to various degrees of **anemia** (up to life-threatening anemia when the concentration of hemoglobin declines very rapidly and/or falls below 0.06 g/L) and (b) that in vitro hemolysis, which is instead caused by inappropriate procedures for collection and/or handling of the biological specimen, can seriously affect patient care and a laboratory's reputation through a wide range of adverse effects on test results.

The objective presence and quantitative measure of the level of hemolysis may be assessed by the measurement of free hemoglobin in plasma samples. The upper reference limit for free hemoglobin in plasma is somewhere between 0.02 and 0.05 g/L for plasma and serum, respectively. Visually, hemolysis is typically established when the free hemoglobin concentrations exceeded 0.3 to 0.6 g/L (18.8 to 37.6 µmol/L), because it confers a detectable pink to red hue to the specimen and becomes clearly visible in samples containing levels of lysed RBCs as low as 0.5% (▶Fig. 3.1). The limit of visual detection might be significantly higher in icteric samples (i.e., those affected by jaundice) [3–5].

In laboratory practice, **hemolyzed specimens** are most appropriately classified according to the concentration of free hemoglobin in serum or plasma, as shown in ▶Tab. 3.1 [3,4].

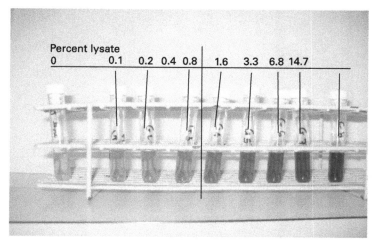

Fig. 3.1: Different degree of hemolysis in hemolyzed specimens.

Tab. 3.1: Classification of hemolyzed specimens

Classification	Free hemoglobin in serum or plasma	Predictable tinge of the specimen
Non-hemolyzed	≤0.05 g/L	Yellow
Slightly hemolyzed	≥0.05 to 0.3 g/L	Yellow to slightly pink
Mildly hemolyzed	≥0.3 to 0.6 g/L	Pink
Frankly hemolyzed	≥0.6 to 3.0 g/L	Slightly red
Grossly hemolyzed	≥3.0 g/L	Red to brown

4 Prevalence of hemolyzed specimens in clinical laboratories

The impact of laboratory testing on **patient care** has gradually increased over the past decades owing to a variety of causes including medical, biochemical, and technological advances; as a result, laboratory test results now contribute to upwards of 70% to 80% of clinical decisions. The widespread introduction of automation, along with innovative tests and technologies – such as molecular biology, proteomics, and nanotechnology – has enormously amplified the diagnostic armamentarium, enabling in vitro diagnostics to be performed with a much greater degree of quality and clinical efficacy. The need to comply with cost-containment policies is forcing laboratory professionals to reorganize several activities of the entire testing process, the considerable improvement in laboratory performance has simultaneously been accompanied by higher expectations from all stakeholders. The traditional perception that key healthcare professionals (e.g., physicians, nurses) had of clinical laboratory testing as a commodity has not changed substantially.

Remarkable advances in test reagents, analytic platforms, and laboratory organization have led to a high degree of quality and efficiency in the analytical phase. This is the phase that is largely under the control of the laboratories performing the tests. Precision, accuracy, and throughput in diagnostics have also improved remarkably. However, progress in the extra-analytical areas of the testing process has been comparatively less manifest. Therefore there is an increasing awareness among the stakeholders that further improvements in global quality and efficacy should now expand beyond the traditional borders of the clinical laboratory, embracing other phases of laboratory performances and addressing the key issues that add value to the healthcare delivery system. Such a new portrayal of the laboratory testing process is now firmly highlighted among the main requirements of modern certification and accreditation programs.

The "magic bullet" to increase quality and efficacy is to provide new paradigm shifts in diagnostic activity, with a major focus on the rationalization, optimization, and standardization of the several extra-analytical processes that convey the greatest potential for quality improvement. This is not a new challenge. The **preanalytical phase** plays a pivotal role among the major determinants of global laboratory efficiency, since the great majority of errors in the entire testing process remain within this area. There are several crucial steps in the preanalytical phase that span factors ranging from the appropriateness of the test request to the correct collection, handling, transportation, and storage of the specimen. The laboratory-nursing interface is probably the most critical issue in this area. As emphasized by Kurec and Wyche, at least five nursing concerns mirror those viewed as important by the laboratory: **quality** issues, laboratory reports, salaries, technical concerns, and professionalism [6]. All these topics are critical and depend mostly on the finite awareness that nurses have regarding the complexity of laboratory diagnostics and the many and varied technical and workflow complications.

Indeed, nurses are relatively unfamiliar with laboratory quality control issues as well, although the continuing introduction of decentralized testing in clinical departments would no doubt increase their familiarity with quality assessment programs and their confidence in them. Nevertheless, nurses' compliance with established and recommended protocols for specimen collection and handling is pivotal. It is this area that may generate a substantial heterogeneity worldwide, yet this essential part of the laboratory testing process is the principal source of unsuitable specimens and usually escapes the direct control or supervision of the laboratory staff [7]. Both personal experience and practical evidence teach us that the clinical staff, including both nurses and physicians, tend to believe that **medical errors** occur from misuse of drugs or mishandled therapies. Thus clinical staff overlook the fact that nonadherence to essential quality criteria for specimen collection and handling would equally threaten their patient's health and consume valuable economic resources owing to the consequent need for sample recollection and/or retesting.

Hemolytic specimens are rather frequently seen in laboratory practice, with an average prevalence described at around 3% of all of routine samples referred to a clinical laboratory and accounting for some 39% to 69% of all the unsuitable specimens received in such laboratories (i.e., nearly five times more than the second major cause) (▶Tab. 4.1) [8]. As widely reported in the literature, in vitro hemolysis also remains the leading cause of the unsuitability of samples from both outpatients and inpatients and both routine and STAT specimens. One of the most cited and fundamental studies was that carried out by the College of American Pathology on 10,709,701 chemistry specimens from 453 participant laboratories. The results demonstrated that as many as 37,208 specimens (0.35%) were rejected prior to testing. The most frequent reason for rejection was hemolysis, which occurred five times more frequently than the second most cited reason, which was insufficient specimen quantity to perform the test [9]. An Italian study involving three large clinical laboratories confirmed hemolysis as the main cause of preanalytical error. In that study, the error rates for hemolysis (as well as other reasons for specimen rejection) were 30-fold higher for inpatients than for outpatients [10]. The most rational explanation for this finding is that **blood-drawing** procedures are performed by more experienced and trained (i.e., laboratory) staff for outpatients, whereas inpatient specimens are collected by more generally trained hospital ward personnel. This, in turn, reflects the more appropriate training and compliance of the former group with the appropriate and recommended procedures for **blood collection** and sample handling. This particular explanation has been confirmed by the results of the study by Gonzalez-Porras et al., who observed a significant reduction in preanalytical errors when tubes for pretransfusion testing were collected directly by blood bank staff rather than clinical staff [11].

Over the past decade, various studies have investigated the prevalence of hemolysis in specimens, and some of these have also assessed the potential burden of adverse events associated with such samples. It is, however, difficult if not impossible to provide a real estimate of in vitro hemolysis for a variety of reasons, including underestimation and underreporting of the problem, huge organizational heterogeneity among different countries and facilities, and the lack of a universal approach for the identification of hemolysis and the identification and reporting of these events.

Over a 30-day observational period, Carraro et al. evaluated the prevalence of hemolyzed specimens received in the STAT section of the Department of Laboratory

Tab. 4.1: Main types of unsuitable specimens arriving in clinical laboratories

Problem	
Absolute prevalence	0.20% to 0.75%
• Inpatients	0.60% to 2.8%
• Outpatients	0.04% to 0.37%
Relative prevalence	
Hemolysis (total)	39% to 69%
• Inpatients	40% to 68%
• Outpatients	18% to 75%
Clotting (total)	5% to 12%
• Inpatients	5% to 15%
• Outpatients	10% to 17%
Insufficient volume (total)	15% to 21%
• Inpatients	10% to 21%
• Outpatients	1% to 13%
Inappropriate container (total)	5% to 13%
• Inpatients	5% to 12%
• Outpatients	8% to 35%
Misidentification (total)	1% to 2%
• Inpatients	1% to 2%
• Outpatients	0.1% to 6%

Medicine of the University Hospital of Padova. Of 27,540 blood specimens from 15,323 sample requests for clinical chemistry, coagulation, and toxicological tests, 505 hemolyzed specimens (3.3%) were identified. Some 64% of these were affected by a small degree (<0.5 g/L of hemoglobin) of hemolysis, 31% by an intermediate degree, and 5% by a high degree (>300 mg/L of hemoglobin). The percentages of hemolyzed specimens received were, however, similar to those received in the internal medicine and surgery departments (3.1%), intensive care unit (ICU) (3.5%), and emergency department (ED) (3.3%) as well as in the organ transplantation unit (3.4%). More importantly, in vivo hemolysis accounted for 16 of 505 cases (3.2%), 7 of which were associated with prolonged extracorporeal circulation during cardiac surgery, 2 with acute ethanol toxicity, 3 with transfusion reactions, 1 with necrotic-hemorrhagic pancreatitis, and 1 with rhabdomyolysis from drug overdose; 2 were of unknown etiology. In 5 of the 16 cases, the presence of hemolysis was not suspected by clinicians, and the laboratory finding was essential in identifying the presence of a critical situation, thus potentially improving the medical outcome in each case [12].

An analysis of the data reported by Romero et al. shows that the percentage of hemolytic specimens in the STAT section of the laboratory of the University Hospital of Malaga was higher in the ED (1.6%) as compared with in-hospital wards (1.0%) and the ICU (0.2%) [13]. In agreement with these findings, Burns and Yoshikawa also observed that the frequency of hemolysis in the laboratory of an acute care teaching hospital of

New York City was significantly higher in the samples coming from the ED as compared with those obtained by expert laboratory phlebotomists (12.4% vs. 1.6%) [14].

A large study carried out from December 2007 to December 2008 in the clinical chemistry laboratory of the Academic Hospital of Verona monitored 150,516 primary tubes for routine and STAT clinical chemistry testing (16,960 from the ED, 2,652 from the hemodialysis unit, 10,116 from the ICU, 62,068 from internal medicine wards, 38,084 from surgical wards, 11,756 from pediatric departments, and 8,880 from the outpatient phlebotomy center), identifying 8,440 hemolytic specimens overall (5.6% of the total). When these hemolytic specimens were classified according to their origin, the highest prevalence was observed for samples referred from the ED (8.8%), followed by pediatric departments (8.5%), internal medicine wards (6.2%), the ICU (5.4%), and surgical departments (4.0%). A lower prevalence was observed for specimens referred from the hemodialysis (1.5%) and outpatient phlebotomy centers (0.1%) [15]. Thus the results of this survey show a high overall prevalence of hemolytic specimens referred for both routine and STAT testing, averaging 5.6% but ranging as high as 8.8%. It was also shown that this prevalence was rather unevenly distributed according to the site of collection, being the highest for samples referred from the ED and the lowest for those collected in the outpatient phlebotomy center, which is the only location where blood collection is typically under the direct control or responsibility of the laboratory. In a further part of this study, aimed to assess the prevalence and type of preanalytical problems for inpatient samples in the coagulation laboratory, the same group analyzed 65,283 routine and STAT test requests for complete first-line coagulation testing (from January 2005 to December 2006). A total of 3,017 preanalytical problems were identified, representing an average of 5.5% of samples, with the highest frequency seen in samples referred from pediatric departments (10.1%). Overall, hemolyzed specimens (19.5%) represented the second most frequent preanalytical problems after samples not received in the laboratory following a test request (49.3%) and were more frequent than samples clotted (14.2%) and samples with inappropriate volume (13.7%). The receipt of hemolyzed specimens was the problem most frequently identified in samples from the ED (the absolute frequency of hemolyzed specimens according to the different wards was reported to be 3.9% in the ED, 1.1% in internal medicine wards, 0.9% in the ICU, 0.8% in pediatric departments, and 0.7% in surgical departments) [16].

In a recent survey promoted by the European Preanalytical Scientific Committee (CEPSC) [17] and the IFCC Working Group "Laboratory Errors and Patient Safety" (388 laboratories worldwide; including 179 from the United States, 188 from Italy, 20 from Australia, 15 from Turkey, and 10 from Czech Republic; comprising 80% hospital laboratories and 20% nonhospital laboratories), the reported prevalence of hemolyzed specimens was as follows (in decreasing order): 39% of laboratories reported from 1% to 3% of all specimens; 28% reported less than 1% of all specimens; 21% reported from 3% to 5% of all specimens; 8% reported from 5% to 10% of all specimens, and 4% reported more than 10% of all specimens. The sources of hemolytic specimens were also reported to be as follows: 53% from emergency units; 16% from pediatric departments; 7% from ICUs, outpatient, and decentralized phlebotomy centers; and 5% from general practitioners and clinical wards (▶Tab. 4.2).

As regards arterial blood gas analysis, Lippi et al. performed a prospective study involving the systematic inspection of all arterial samples referred for testing. Hemolysis

Tab. 4.2: Results of an international survey on hemolysis

Query	Answer
Are you aware of the problem that some lab tests might be affected by hemolysis?	Yes: 91% No: 9%
How do you check to evaluate sample hemolysis in your lab?	Measure the hemolysis index: 43% Visual inspection: 56% I do not check: 1%
Do you systematically monitor the number and origin (wards, facilities, etc.) of hemolyzed specimens that you receive in your lab?	Yes: 58% No: 42%
What is the percentage (approximately) of hemolyzed specimen you receive?	Less than 1%: 28% 1% to 3%: 39% 3% to 5%: 21% 5% to10%: 8% 10% to 20%: 2% Greater than 20%: 2%
Where do most hemolyzed specimens come from?	Emergency unit: 53% Pediatric department: 16% Intensive care unit: 7% Outpatients: 7% Decentralized phlebotomy centers: 7% General practitioners: 5% Clinical wards: 5%
How do you deal with hemolyzed specimens?	Perform all requested tests but do not report the test results mostly affected by hemolysis: 44% Reject the specimen, refuse to perform requested tests in cases of hemolysis: 56%

was assessed immediately after arterial blood gas analysis was completed by transferring the blood into secondary tubes and centrifuging these at 3,500 g for 5 minutes. Of a total of 1,228 specimens received in the laboratory during the 1-month study period, 15 specimens (i.e., 1.2%) with various degrees of hemolysis were identified, the vast majority having been referred by the nephrology, dialysis, and transplantation unit (12 of 15, or 80%) [18].

5 In vivo hemolysis

5.1 Causes

As noted previously, the life span of RBCs may be shortened in a number of disorders, often resulting in anemia of varying degrees if the bone marrow is not able to adequately replenish the prematurely destroyed cells. The disorders associated with **hemolytic anemias** are generally identified by the abnormality that brings about the premature destruction of the RBCs (e.g., sickle cells are characterized by a short and highly variable survival, so that only 30% might remain after 6 to 16 days) [1]. The RBCs may be prematurely removed from the circulation by macrophages, particularly those of the spleen and liver (extravascular hemolysis) or, less commonly, by the disruption of their membranes during their circulation (intravascular hemolysis).

Anemia caused by in vivo hemolysis is generally called hemolytic anemia, which represents approximately 5% of all anemias. Hemolytic anemia has numerous possible causes and several plausible consequences, ranging from relatively harmless to life-threatening [19–38]. A conventional classification of hemolytic anemias encompasses both inherited and acquired causes (▶Tab. 5.1). Among inherited disorders causing hemolytic anemias, the most frequent are those related to defects in hemoglobin production (i.e., hemoglobinopathies; sickle cell disease and the thalassemias being the most frequently encountered), defects of RBC membrane production (hereditary spherocytosis being the commonest, with an incidence of approximately 1:1,000 to 1:4,500 of all newborns), and defective red cell metabolism (especially defects in enzymes involved in the Embden-Meyerhof pathway and in the exose-monophosphate shunt, by far the most common being glucose-6-phosphate dehydrogenase [G6PD] deficiency, affecting more than 200 million people throughout the world). It must be noted that both hemoglobin S (the abnormal hemoglobin causing sickle cell disease) and G6PD deficiency probably exert protective effects from malaria by providing a defective home for the merozoite.

In most patients with acquired hemolytic anemia, RBCs are normally produced but prematurely destroyed because of damage acquired in the circulation. The rare disorders characterized by acquired dysplasia of the bone marrow cells and the production of structurally and functionally abnormal red cells constitutes an exception to the norm. The RBC damage that occurs may be mediated by antibodies or toxins that mark the cells for premature death, or it may be due to other problems encountered during their passage through the circulation, including an overactive mononuclear phagocyte system or traumatic lysis by natural or artificial impediments to smooth circulation. The acquired hemolytic anemias can be classified into five categories: (a) entrapment (hypersplenism); (b) immune; (c) traumatic; (d) due to toxic effects on the membrane; and (e) paroxysmal nocturnal hemoglobinuria (PNH).

The spleen is particularly efficient in trapping and destroying RBCs that have minimal defects, often so mild as to be otherwise undetectable by in vitro techniques. This unique ability of the spleen to filter mildly damaged RBCs results largely from its unusual

Tab. 5.1: Leading causes of in vivo hemolysis

Inherited hemolytic anemias
- Defects in hemoglobin production
  - Thalassemias
  - Sickle cell disease
- Defects of RBC membrane production
  - Hereditary spherocytosis
  - Hereditary elliptocytosis
  - Paroxysmal nocturnal hemoglobinuria (PNH)
- Defective red cell metabolism
- Glucose-6-phosphate dehydrogenase deficiency
- Pyruvate kinase deficiency

Acquired hemolytic anemias
- Immune-mediated causes
  - *Mycoplasma pneumoniae* infection (cold agglutinin disease)
  - Autoimmune hemolytic anemia (AIHA)
  - Autoimmune diseases (systemic lupus erythematosus and chronic lymphocytic leukemia)
- Hypersplenism
- Burns
- Infections
  - Malaria
  - Babesiosis
  - *Clostridium*
- Mechanical damage in circulation
  - Disseminated intravascular coagulation (DIC)
  - Hemolytic uremic syndrome (HUS)
  - Thrombotic thrombocytopenic purpura (TTP)
  - Prosthetic cardiac valves
  - HELLP (hemolysis, elevated liver enzymes, and low platelets) syndrome
- Transfusion of blood from a donor with a different blood type
- Drugs, toxins and other miscellaneous causes

vascular anatomy, in particular the presence of the sinuses in the spleen's red pulp. Diseases causing splenomegaly are also capable of causing hemolytic anemia. Myeloproliferative disorders, lymphomas, storage diseases (e.g., Gaucher's disease), as well as systemic inflammatory diseases leading to splenic hypertrophy or diseases that cause congestive splenomegaly (e.g., liver cirrosis or thrombosis of the portal, hepatic, or splenic veins) all can cause various degrees of hemolytic anemia. Autoimmune hemolytic anemia (AIHA) may result from warm (i.e., reacting at body temperature) or cold (i.e., reacting at temperatures below 37°C) autoantibody types or, rarely, from mixed types. Burns, infections, drugs, toxins, mechanical damage during circulation, transfusion of blood from a donor with a different blood type, and other miscellaneous causes can also lead to AIHA. The AIHA of the warm antibody type is induced by IgG antibodies and occurs at all ages but is more common in adults, particularly women. In approximately one fourth of patients, this disorder occurs as a complication of an underlying disease affecting the immune system, especially neoplasms of the immune system (e.g., chronic lymphocytic leukemia, non-Hodgkin's lymphoma, and Hodgkin's lymphoma) and collagen vascular diseases, especially lupus erythematosus. The severity of the anemia can range from mild to the fulminant, overwhelming, and often fatal hemolysis associated with hemoglobinemia, hemoglobinuria, and shock. Cold autoantibodies (also called cold agglutinins) arise in two clinical settings: (a) monoclonal antibodies as the product of lymphocytic neoplasia (rarely nonlymphocytic neoplasia) and (b) polyclonal antibodies produced in response to infection. Transient cold agglutinins occur commonly in two infections: *Mycoplasma pneumoniae* infection and infectious mononucleosis. In both the titer of antibody is usually too low to result in clinical symptoms, but its presence is of diagnostic value; only occasionally is hemolysis present. The clinical manifestation elicited by the cold agglutinins upon exposure to cold are of two sorts: intravascular agglutination (acrocyanosis: the marked purpling of the extremities, ears, and nose when the blood becomes cold enough to agglutinate in the veins) and hemolysis, usually not severe.

Drug-induced hemolytic anemia is typically classified according to three mechanisms of action: drug absorption, immune complex, or autoantibody. These IgG- and IgM-mediated disorders produce a positive direct antiglobulin test (DAT; also known as the direct Coombs test) and are clinically and serologically distinct from AIHA. Hemolysis due to high-dose penicillin is a typical example of the drug-absorption mechanism (anti–IgG positive), in which a therapeutic compound attached to the RBC membrane triggers the production of (class IgG) antibodies. When large amounts of drug coat the cell surface, the antibody binds the cell membrane and causes hemolysis. Quinine-induced hemolysis is instead the prototype of the immune complex mechanism, in which the drug induces the production of (class IgM) antibody. The drug-antibody complex binds to the RBC membrane, promotes activation of complement, and thereby causes hemolysis (anti–IgM positive). Finally, the action of the drug alpha-methyldopa is a typical example of anti–RBC antibody induction. The exact mechanism is only partially known, this drug can promote the production of anti–RBC IgG antibodies and thus to cause hemolysis, probably by altering an RBC membrane protein and rendering it antigenic (anti–C3 positive).

RBCs may be fragmented by mechanical trauma as they circulate; this invariably leads to intravascular hemolysis and in most cases to red cell fragments called schistocytes (identified by the sharp points that result from faulty resealing of the fractured membrane). Traumatic hemolysis occurs in three clinical settings: (a) when RBCs flow through small vessels over the surface of bony prominences and are subject to external impact during various physical activities; (b) when RBCs flow across a pressure gradient created by an abnormal heart valve or valve prosthesis (most frequently an aortic valve prosthesis) and are disrupted by a shear stress (macrovascular); and (c) when the deposition of fibrin in the microvasculature exposes them to a physical impediment that fragments them (thrombotic thrombocytopenic purpura, hemolytic-uremic syndrome, disseminated cancer, hemangiomas, and eclampsia being the most frequently encountered).

Exercise-induced hemolysis has been reported for more than 50 years. Distance running has been particularly associated with significant destruction of RBCs, with erythrocyte turnover being substantially higher in runners compared with untrained controls. The most reasonable explanation is the mechanical damage sustained by RBCs as they pass through the capillaries of the foot during the footstrike phase. Additional studies on athletes involved in sports in which foot impact does not occur (e.g., swimming, cycling, weight lifting, and rowing) have, however, found evidence of some degree of exercise-induced hemolysis. Besides footstrike, several other mechanisms may therefore contribute to hemolysis during exercise, including continuous exposure to high oxygen flux, which causes oxidative damage to RBCs; perturbation of osmotic homeostasis, which might render the erythrocytes more susceptible to membrane damage; and hemolysis, especially during their passage through the microcirculation, where cells undergo a constant swelling and shrinking cycle. Finally, it is also possible that the effect on capillaries of the compression of large muscle groups may accelerate the hemolysis of older RBCs [39,40]. As in the case of strenuous physical exercise, activities involving repeated mechanical trauma might also trigger in vivo hemolysis. Such activities would include hand drumming (e.g., extracorpuscular hemolysis was found in up to 20% of Candomblè drummers in Uruguay, owing to multiple manual traumas) [41], basque ball [42], rugby, and judo.

Last but not least, PNH is a distinctive form of hemolytic disorder because it is an intracorpuscular defect acquired at the stem cell level [43]. PNH is an acquired clonal disease, probably arising from an inactivating somatic mutation in a single abnormal stem cell of a gene involved in the biosynthesis of the glycosylphosphatidylinositol anchor, leading to a defect in at least 20 proteins of the RBC membrane.

5.2 Clinical presentation

The clinical presentation of in vivo hemolysis depends on whether the onset of the disorder is gradual or abrupt as well as on the severity of RBC destruction. Patients with mild hemolysis may be totally asymptomatic, while more serious forms of anemia can be life-threatening. The symptoms of hemolytic anemia are similar to those of other forms of anemia (e.g., fatigue, pallor, shortness of breath), but the breakdown of RBCs also leads to icterus and jaundice and even to a red-brown urine (hemoglobinuria). It also increases the risk of particular long-term complications, such as splenomegaly, encountered in a variety of hemolytic anemias, which is itself a worsening factor for the hemolysis (hypersplenism). Chronic hemolysis, in particular, can lead to an increased excretion of bilirubin into the biliary tract, thereby predisposing to the development of gallstones. The continuous release of free hemoglobin has also been associated with the development of pulmonary hypertension, revealing itself with episodes of syncope, chest pain, and progressive breathlessness. In conditions where the rate of RBC breakdown is increased, the body initially compensates by producing more RBCs (and reticulocytes); however, the breakdown of RBCs can occur at a rate exceeding the body's ability to make new RBCs, and thus anemia can develop. The cardiac signs of anemia (flow murmurs, edema, etc.) may also be present. The global incidence of death in cases of hemolytic anemia is low, older patients and patients with cardiovascular imbalance are at increased risk. Morbidity is mainly related to the etiology of the hemolysis and the underlying disorder (such as sickle cell anemia or malaria).

5.3 Diagnosis of hemolytic anemia

Hemolysis is typically associated with a release of intracellular components from the RBCs, especially hemoglobin, lactate dehydrogenase (LDH), aspartate aminotransferase (AST), and potassium (▶Fig. 5.1). During in vivo hemolysis, the rise in LDH (which also causes a shift in the LDH electrophoresis pattern, particularly of LDH1 and LDH2) occurs when the reticulocyte index is above 10%. In subjects with an LDH activity of 165 U/L and an in vitro hemolysis of 0.8 g/L, the serum hemoglobin causes a nearly 50% increase in LDH activity. Another hallmark of hemolytic anemia is the variable increase of immature RBCs (reticulocytes), this usually occurs with a delay of 24 to 48 hours. An increase in indirect bilirubin and urobilinogen is also evident owing to the hemoglobin released into the plasma. Bilirubin, a breakdown product of hemoglobin, can accumulate in the blood, causing jaundice, and can also be excreted in the urine, causing it to turn dark brown. In the presence of hemolytic anemia due to a mechanical cause, fragmented RBCs (schistocytes) are visible in the peripheral blood smear. In patients with AIHA, most warm autoantibodies are immunoglobulin (Ig) G, and these can be detected with the direct Coombs test or DAT test. Intravascular hemolysis accompanied by a decrease in the hematocrit of more than 25% within 12 hours can also cause hypertriglyceridemia due to the reduced microcirculation and/or mobilization of free fatty acids and their reesterification to triglycerides.

In blood, haptoglobin acts by binding free hemoglobin molecules, released from RBCs, with high affinity and therefore inhibits the oxidative activity of hemoglobin. The haptoglobin-hemoglobin complexes are then removed from the circulation by the reticuloendothelial system, mostly in the spleen. In clinical settings, the haptoglobin assay is hence useful to screen for and monitor intravascular hemolysis, since free hemoglobin released after RBC breakdown is bound to haptoglobin, causing a significant reduction in the levels of this protein. On the contrary, in the setting of extravascular hemolysis, the reticuloendothelial system internalizes the erythrocytes directly

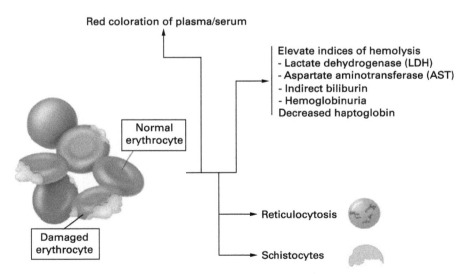

Fig. 5.1: Biochemical and hematological consequences of hemolysis.

and haptoglobin levels are normal. Therefore a low serum haptoglobin level is a basic criterion for diagnosing moderate to severe intravascular hemolysis. Nevertheless, haptoglobin is an acute-phase reactant and the presence of concomitant infection, other reactive states, or chronic hemolysis may mask the diagnosis by raising haptoglobin levels.

Finally, injury and breakdown of platelets and leukocytes can also influence test results without the presence of visual hemolysis. It is, in fact, clear that thrombocytolysis is responsible for the higher concentration of a number of intracellular components in serum compared with the plasma. The intravascular destruction of leukocytes can lead to increased lysozyme levels in myeloid and monocytic leukemias.

6 In vitro hemolysis

Despite recent advances in technology, which have contributed to limit the probability and the adverse clinical outcome of errors in the analytical phase of testing, a large degree of variability still occurs for individual or system design defects in the extra-analytical phases of the total testing process. This applies especially to the pre-analytical setting, which incidentally comprises the most labor-intensive and high-risk activities [7,8,44–46]. Preanalytical problems, which still account for nearly 70% of total errors encountered within the total diagnostic process, can produce doubtful or spurious results, and they consistently appear as cross-boundary issues that may affect all stakeholders of laboratory testing by consuming healthcare resources and compromising patient care. The most common preanalytical problems are traditionally identified as inadequate procedures for collection, including inappropriate quality of the specimen (hemolysis, clotting, contamination), insufficient volume, inappropriate containers, and misidentification. Although a variety of congenital and acquired disorders can lead to in vivo hemolysis (e.g., hemolytic anemia, as in Chapter 5, Section 1), there are also several conditions that can cause in vitro (spurious) breakdown of RBCs during the collection, handling, and storage of blood specimens. The factors producing in vitro hemolysis can therefore begin at the patient's bedside and continue throughout the total testing process, up to analysis of the sample. The responsible factors can be classified as dependent upon the patient's condition (e.g., fragile veins), the skill of the person collecting the sample (e.g., training), as well as the conditions for transport, processing, and storage of the specimens [3,4].

Regardless of the various potential causes listed in the previous paragraph and in ▶Tab. 6.1, the leading sources of hemolytic specimens involve improper procedures for their collection and handling. Whereas most of these procedures are usually outside the control of the laboratory, unsuitable samples and inaccurate test results are often incorrectly attributed to laboratory errors.

6.1 Unsuitable procedures during blood collection

Blood cells in general and RBCs in particular are fragile and susceptible to rupture when exposed to mechanical trauma, osmotic shock (exposure to nonisotonic fluids), and temperature extremes (e.g., during transportation to the laboratory). Consider also that if a vein is traumatized during puncture, the first tube collected may contain hemolyzed blood, while subsequent tubes might be fine. Excessive aspiration can cause the RBCs to be literally smashed on their way through the hypodermic needle owing to turbulence and physical forces. This type of hemolysis is more likely to occur when a patient's veins are difficult to find, when they collapse when blood is removed by a syringe or a modern vacuum tube, or when excessive tournique pressure is maintained during collection. Accordingly, experience and use of proper technique by the phlebotomist or nurse are key to the prevention of hemolysis. Alternative sites to the

Tab. 6.1: Leading causes of in vitro hemolysis

- Patient-dependent
  - Fragile veins
  - Unsuitable venous access
- Operator-dependent
  - Skill of the operator
  - Location of needlestick
  - Traumatic blood draw
  - Unsatisfactory attempts
  - Missing the vein during venipuncture
  - Drawing from a hematoma
  - Capillary collection
  - Antiseptic used for phlebotomy
  - Prolonged tourniquet placement
  - Fist clenching
  - Tube underfilling
- Device-dependent
  - Collection with unsuitable devices
  - Syringe
  - Catheters and intravenous (IV) access
  - Butterfly devices
  - Small-gauge needles
  - Partial obstruction of IV lines
  - Use of the handheld lance, especially for pediatric blood collection
- Handling of the specimen
  - No mixing or insufficient mixing
  - Excessive shaking or mixing
  - Syringe transfer
- Transport of the specimen
  - Origin of the specimen (maternity, emergency, and intensive care departments)
  - Transport modality (pneumatic tube, courier)
  - Transport conditions (mechanical trauma, time, temperature, and humidity)
  - Direct contact of the tubes with ice or frozen gel packs
- Sample processing
  - Time delay before centrifugation
  - Centrifuge condition (force, time, temperature)
  - Poor barrier integrity
  - Specimen respun
- Storage of the specimen
  - Respin add-on
  - Condition of storage (temperature and duration)

antecubital area, such as hand veins, are fragile and easily traumatized; therefore collection of blood from these sites is more likely to produce spurious hemolysis. Performing the **venipuncture** before the alcohol has thoroughly dried may also cause RBC rupture [47]. Use of the wrong type of tube can result in hemolysis due to chemical or physical actions, especially in some physiopathological situations. Partial underfilling, which might occur rather frequently while drawing blood, is often overlooked as a potential cause of hemolysis. Tamechika et al. compared complete filling with one-fifth filling of the primary tube and showed that the former circumstance produces on average elevations of LDH, AST, and potassium levels of 8.0%, 3.8%, and 3.4%, respectively (all p <0.01) [48]. The presence of microhemolysis was also confirmed using a urine stick method; this has been mainly observed when vacuum sample containers were incompletely filled or when centrifugation of the specimens was substantially delayed.

Venous stasis is frequently overlooked as a potential source of preanalytical variability in test results, it is instead accompanied by multiple **interferences** in laboratory testing as well as by a greater likelihood of generating spurious hemolysis. As specifically regarding to the use of the tourniquet, an excessively prolonged placing (e.g., over 3 to 5 minutes) – and the resulting venous stasis – promotes the exit of water, diffusible ions, and low-molecular-weight substances from the vessels, thereby increasing the concentration of various blood analytes at the punctured site and influencing the reliability of test results. Moreover, when the vascular microenvironment is subjected to both hypoxia and concurrent stasis, there is a good chance that bioproducts such as protons will accumulate, and these have the potential to promote changes in laboratory parameters [49–51]. Regardless of these important changes, a greater degree of hemolysis has been noticed in samples collected with prolonged tourniquet placement [52]. Saleem et al. carried out a prospective study of blood sampling events to obtain detailed information on the potential causes of hemolysis [53]. Hemolysis in the specimens was systematically assessed by measuring the **hemolysis index** (HI) on a Roche modular system and a threshold corresponding to 48 g/L of free hemoglobin in serum was used to classify the samples as hemolyzed. The overall incidence of hemolyzed specimens among the samples studied was 6.5%. Analysis of the results showed statistically significant univariate associations between hemolysis and the staff group, tourniquet time, and number of attempts at venipuncture. Nevertheless, the final logistic regression model included only tourniquet time (p <0.001), with an increased risk of hemolysis corresponding to increased tourniquet time (the odds ratio for hemolysis when the tourniquet time exceed 1 minute was 19.5; 95% CI 5.6–67.4%). Different results were instead reported by Serdar et al., who, although not measuring the content of free hemoglobin in the samples directly, could observe only modest variations in the indirect markers of hemolysis (e.g., 1.9% for potassium, 4.3% for LDH, and 3.3% for AST, respectively) as a function of tourniquet placement (up to 60 seconds) [54]. The reasons for hemolysis during prolonged venous stasis are mostly unknown, it has been speculated that hemoconcentration and altered water balance in the cells might occur, thereby causing lysis of RBCs and platelets. When the tourniquet is left in place too long, a hematoma can develop as well [55]. Repeated fist clenching with or without application of the tourniquet might also cause damage to the RBCs and thereby result in a hemolyzed blood specimen.

6.2 Blood collection with various devices and needles

The evacuated tube collection system consists of a double-pointed needle, a plastic holder or adapter, and a series of vacuum tubes with rubber stoppers (▶Fig. 6.1). The procedure of collecting blood with these devices seems to produce the best-quality blood samples for laboratory testing and contextually ensures a greater degree of safety for the operators because the blood flows directly into the appropriate blood collection tubes with no external contact or requirement for additional syringes. Moreover, the sheath makes it possible to draw several tubes of blood by preventing the leakage of blood as tubes are changed. One of the leading issues involving the problem of hemolyzed specimens is collection with inappropriate or unsuitable devices, such as intravenous (IV) catheters, butterfly devices, and small-bore needles.

6.2.1 Blood collection by syringe

The leading advantage of using a syringe for a blood draw is that it affords the operator the ability to control the vacuum or pressure being applied inside the vein. Nevertheless, the use of a syringe rather than a vacuum tube system for collecting blood is now strongly discouraged for a variety of economic (e.g., additional cost of the syringes) and safety reasons (e.g., the high risk of needlestick injury and subsequent infection in transferring blood from the syringe to the blood collection tube). Moreover, in a study performed by Becton Dickinson Diagnostics, visual hemolysis was observed in 19% of specimens obtained with a syringe as compared with only 3% with an evacuated tube system [56]. These data have been confirmed in other studies. Carraro et al. showed that the vast majority of hemolyzed specimens were due to blood being drawn too vigorously through needles into the syringe (30.7% of cases), butterfly needles into the syringe (20.0% of cases), IV catheters into the syringe (16.5% of cases), and infusion access into the syringe (11.5% of cases) [12]. Grant also observed that the samples were more likely to be hemolyzed when they were collected with a syringe rather than with an evacuated tube system (9% vs. 3%) [57].

Basically the transfer of blood from a syringe to a primary blood tube can trigger RBC injury and rupture, since excessive force applied to the plunger during blood collection causes the blood to enter the tube forcefully. Blood can also clot and hemolyze while being aspirated into a large-volume syringe, and the forceful transfer of blood from the syringe into the blood collection tube can damage the RBC membrane. Accordingly,

The syringe body has a syringe top for receiving a seal assembly wherein it is reversibly locked to the top of the syringe.

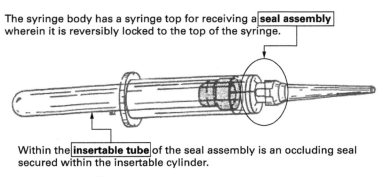

Within the insertable tube of the seal assembly is an occluding seal secured within the insertable cylinder.

Fig. 6.1: Evacuated tube collection system.

hemolysis of 2- and 9-days stored sedimented RBCs infused by a constant-rate syringe delivery pump through a 25-gauge needle has been clearly described by Wilcox et al. [58].

6.2.2 Blood collection from intravenous catheters

Several types of patients who receive invasive medical treatment or are subjected to diagnostic investigations undergo subcutaneous venous cannulation by permanent devices. The venous cannula may be kept in place for a few minutes or for hours (e.g., in patients undergoing general anesthesia, sedation before surgery, noxious clinical procedures, or diagnostic radiological investigations), days, and even weeks (e.g., hemodialyzed, critically ill or cancer patients undergoing long-term chemotherapy). Blood products, fluids, electrolytes, antimicrobial drugs, long-term infusions for pain relief, or chemotherapy and the infusion of other essential therapeutic agents can be delivered efficiently using these devices. In these circumstances, for the patient's well-being, it is nevertheless essential to maintain catheter patency. Concurrently, a second venipuncture to draw blood for laboratory testing might be inopportune or inconvenient because the same venous access can theoretically be used. This latter approach is also cheaper, as no other devices for blood collection are required, and it might even be much safer for both the patient and the phlebotomist. Therefore nurses occasionally perform phlebotomy via IV catheters (▶Fig. 6.2a) so as to improve efficiency in short-stay or procedural units and to reduce patient discomfort from double punctures (one from placing the catheter and the other from collecting blood).

It is now well established, however, that specimens obtained from peripheral IV catheters are frequently underfilled (the large air space in these collection systems leads to underfilling of the first blood collection tube) or hemolyzed, and this practice can even dislodge catheters and necessitate reinsertions. All these unwelcome occurrences can ultimately lead to multiple needlesticks and delayed treatment, thus reducing patients' satisfaction and also raising the cost of their care substantially. The leading mechanism responsible for damage to RBCs during blood collection with these devices is shear-induced hemolysis due to biaxial tension in excess of that necessary to produce a critical area strain for a certain time. Above this threshold, pores in the cell membrane allow cell contents to escape or the cell membrane to fragment. After brief exposure to high stress, pores may reseal before or after cell contents have escaped, producing ghost cells in the latter scenario. An additional cause of hemolysis is blood flowing through different internal diameters (catheter and connectors) and at various angles; accordingly, the resultant turbulence can cause RBC rupture [59].

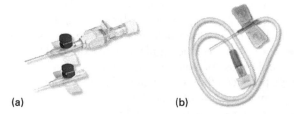

(a)　　　　　　　　　　　　(b)

Fig. 6.2: Intravenous catheter (a) and butterfly needle device (b).

It has previously been reported that the prevalence of hemolyzed specimens might range from 15% to 25% when catheter diameters from 22- to 20-gauge are employed as compared with 3.8% in samples collected by a 21-gauge conventional straight needle [60].

A recent meta-analysis retrieved eight observational, descriptive, comparative, and experimental studies to compare in vitro hemolysis between specimens collected from IV catheters and those collected from regular venipunctures (in most of these studies hemolysis was assessed via visual inspection to detect a color change indicating the presence of hemoglobin; in a minority of studies, hemolysis was measured with spectrophotometers) [61]. The results of these studies showed that hemolysis varied considerably between methods and units, occurring in 3.3% to 77% of blood samples obtained via IV catheters, whereas it occurred in only 0% to 3.8% of those obtained via regular venipuncture. Hemolysis of samples obtained from IV catheters via different methods also ranged from 5.6% to 77% for samples obtained with a vacuum tube, 3.3% to 49% for samples obtained with a syringe, 12.8% to 49% for samples obtained from new IV catheters, and 24% for samples obtained from established IV catheters. Factors associated with in vitro hemolysis could be classified as anatomical/physiological, equipment-related, or technical. The first category included the right arm (e.g., obtaining samples from the dominant side of the patient's body might be easier), forearm, or antecubital space, smaller distal veins and − surprisingly − discharge diagnoses (respiratory, gastrointestinal, reproductive, dermatological, endocrine). It was also highlighted that the use of distal veins can also lead to more hemolysis compared with the antecubital fossa because of the smaller diameters, slower flow, and increased resistance. Factors related to equipment included plastic, smaller, and new IV catheters; use of vacuum tubes or syringes; partial catheter obstructions; and primary tube size. Partial catheter obstructions were deemed to increase aspiration force when syringes were used to collect samples. Among technical factors, the list included difficult catheter placements or cumbersome collection, multiple attempts at the placement of IV catheters, tubes filled to less than half full capacity, and excessive force for aspirating blood or filling tubes. Prolonged tourniquet time was an additional cause of spurious hemolysis owing to increases in venous pressure and extravasation of fluid into the intracellular space. Therefore the reported studies and manufacturers' information led the authors to conclude, with class III evidence (not acceptable or useful; may be harmful), that the collection of laboratory specimens from peripheral IV catheters was not to be recommended. In agreement with these considerations, a study of Burns and Yoshikawa [14] identified the widespread practice of collecting blood from plastic catheters connected to patients (aimed to reduce the number of venipunctures) as the leading problem causing hemolysis in the ED. In particular, a comprehensive analysis of the data demonstrated that the leading factors contributing to spurious hemolysis included drawing blood from a distal arm versus the antecubital fossa ($p = 0.0054$), the use of a small (i.e., 22-gauge) versus a bigger (i.e., 20-gauge) plastic cannula ($p = 0.010$), the collection of less than half of a full tube versus greater than or equal to half of a full tube ($p = 0.016$), tourniquet placement extended to >2 minutes versus ≤2 minutes ($p = 0.016$), and the use of a plastic versus a metal cannula ($p = 0.016$). Collection of blood by a syringe produced the same degree of hemolysis as that observed by using an evacuated tube system ($p = 1.00$). Blood collection by syringe and transferral of blood into primary tubes is discouraged, owing to the inherent biological risk of incurring needlestick injuries, the low degree of hemolysis associated with this procedure is not surprising inasmuch as it would prevent the well-known risk of mechanical break-down of RBCs due to the combination of an evacuated blood collection system with a cannula. The latter procedure is commonplace in EDs in order to limit the number of ve-nipunctures (e.g., the combination of vaccum tubes and cannulas determines hemolysis due to excessive aspiration force of the vacuum through the disposal). In another study, Ong et al. found that the use of vacuum tubes was associated with the highest chance of producing spurious hemolysis, while additional important causes were use of IV dispo-sals for blood sampling rather than repeated venipunctures, reduced sample volume, and prolonged interval from sampling to analysis [62]. Lowe et al. carried out a prospec-tive crossover study of blood collection techniques in a large community teaching hos-pital ED, and reported that the rate of hemolysis was much (and significantly) higher in IV catheter samples (5.6%) versus venipuncture samples (0.3%; p <0.05) [63]. Similar fig-ures were reported by Kennedy et al. (13.7% hemolysis in samples obtained through the IV catheter vs. 3.8% in those obtained with separate vacuum tube venipuncture), and Grant (20% hemolysis and test cancellation in blood drawn through IV cathe-ters vs. <1% in blood samples collected by straight needle) [57]. In a separate study aimed to identifying the factors related to hemolysis rates in blood samples drawn from IV sites in the ED, the leading causes of spurious hemolysis included IV placement sites in the right hand or forearm; antecubital placement and use of a small-caliber IV catheter (i.e., <22-gauge), blood drawing categorized as difficult, number of tries for IV placement, small blood tube size, and discharge diagnoses of respiratory, gastrointes-tinal, reproductive, dermatologic, and endocrine diseases [64]. A critical review was carried out to pool results of observational, descriptive, comparative, and experimental studies comparing hemolysis between specimens collected from IV catheters and those collected from venipunctures in the ED. The rate of hemolysis varied considerably among the different studies, but hemolysis occurred at a remarkably lower rate when blood samples were collected via venipuncture (0% to 3.8%) as compared with those obtained via IV catheters (3.3% to 77%) [61]. The leading factors associated with hemolysis were (a) anatomical – venipuncture in the right hand, forearm, antecu-bital space, or smaller distal veins – and physiological, including discharge diagnoses (respiratory, gastrointestinal, reproductive, dermatological, endocrine disease); (b) equipment factors – plastic, smaller, and new IV catheters, partial catheter obstructions, laboratory tube of larger size; and (c) technical factors – difficult catheter placements, difficulty collecting blood, multiple or unsuccessful attempts to place IV catheters, par-tial filling of the primary vacuum tubes and excessive force in aspirating blood or filling tubes with a syringe.

In the face of all this evidence, it is suggested that blood samples not be collected when new IV catheters are started or from established IV catheters. Some notable ex-ceptions might include instances where patients are receiving thrombolytic agents or are at increased risk of bleeding, or possibly in an emergency situation where there is limited vascular access, even though hemolysis may still result and delay critical treatment.

6.2.3 Blood collection from butterfly devices

The butterfly device is a small needle attached to flexible plastic wings that is con-nected to flexible extension tubing (▶Fig. 6.2b). These disposable devices are often regarded as a reliable alternative to the classic straight needle for the collection of blood in selected categories of patients because the adapter can easily be connected to the disposable butterfly (thus it would fit both into a vacuum method needle holder and a traditional vacuum system). Therefore blood collection using a butterfly system is easier and less painful in dealing with newborns, children, small animals, and patients with small, difficult, and atypical venous access, where it might be very difficult to rest the supporting hand (venipunctures in hand, leg, heel, or cranium). The use of a butterfly device, which is less intimidating because of the smaller needle, is also advocated in other circumstances, especially in approaching nervous or anxious patients or when a large number of blood tube collections are required. Its simplicity of use is another major advantage for unskilled or nervous operators, since the needle does not have to be held still once it is in the vein. Practical disadvantages are, however, represented by the additional cost of the devices, the greater chance of needlestick injury, and the possibility that a small amount of blood will be spilled when the needle is withdrawn from the skin. Such a spill may stain collection furniture or bedding and also carries an infection risk. The use of these systems for drawing blood specimens has also been discouraged in laboratory practice unless more conventional approaches have failed, in part for reasons of cost but largely for the greater likelihood of obtaining unsuitable samples (incomplete filling of the vacuum tube, hemolysis, activated or clotted samples). In a previous study, we compared results of hematological and clinical chemistry testing after drawing blood into evacuated tubes, employing either a traditional straight needle or a butterfly device with 300-mm-grade polyvinyl chloride (PVC) tubing [65]. Blood samples and complete sets of data were successfully obtained for 30 consecutive outpatients. Of the 43 hematological and clinical chemistry parameters measured, means for paired samples collected by the two alternative drawing techniques did not differ significantly except for a few analytes. We observed a significant increase (+1.7%) in potassium using the butterfly devices, the degree of hemolysis (as assessed by the measurement of free plasma hemoglobin, LDH, and AST) in blood specimens collected via a 300-mm PVC tubing device appeared not to be greater than that of specimens obtained with a traditional straight needle. This result is consistent with the study of Sonntag [66], who showed that the concentrations of plasma potassium and LDH are significantly affected by hemoglobin concentrations >0.2 g/L. The 95% agreement interval in the set of differences was also acceptable and mostly corresponded with the current analytical quality specifications for desirable bias of all analytes tested. Taken together, the results of this investigation would suggest that, when proper technique is used within certain limitations and, at differently from IV catheters, the butterfly device may be a reliable alternative to the conventional straight needle to draw blood for laboratory testing [65]. However, care must be taken in regard to underfilling of the first blood collection tube owing to the large air space in the cannula tubing; the typical approach is to discard the first tube collected with such devices [67].

6.2.4 Blood collection with small-bore needles

There are some general characteristics that typically identify size and purpose for phlebotomy needles. The needles are traditionally calibrated by gauge (G), which refers to the diameter in millimeters; the larger the G number, the smaller the diameter of the bore. Blood collection is usually performed with needles ranging from 19 to 25 G.

Nineteen to 21 G needles are used primarily for large antecubital veins, 23 G needles for smaller antecubitals, medium size forearm, hand veins and 25-gauge or smaller needles are used for the smallest veins or in newborns and children. Small needles might be also suitable for nervous or anxious patients, who suffer from anticipatory fear of needle insertion, or "needle phobia." There are no definitive guidelines on the needle bore size to be used for collecting venous blood samples, it is widely accepted that blood must be withdrawn with a suitable needle to avoid excessive pressure or shear stress, which might finally lead to damage or rupture of blood cells, especially erythrocytes, producing unsuitable samples due to the presence of clots or hemolysis. Therefore, the use of small size needles for venipuncture has been for long discouraged, on the basis of some evidences that needles equal to or smaller than 23 gauge might introduce a high degree of preanalytical variability, that might ultimately influence the reliability of test results. In a study, Lippi et al. investigated the influence of the needle bore size used for collecting venous blood samples on routine clinical chemistry testing, including free hemoglobin in plasma. Not surprisingly, the concentration of free plasma hemoglobin in samples collected with a <23-gauge needle was substantially increased when compared with a traditional 21-gauge needle (+1.3%, $p < 0.001$). Moreover, when expressing the bias of results in terms of percentage of the mean for paired samples, the variation of free plasma hemoglobin concentration in samples collected with a <23-gauge needle largely exceeded the desirable bias of the analyte (6.8% versus ±1.8%). The results of this investigation suggest therefore that <23-gauge needles, cannot be universally recommended when collecting venous blood for clinical chemistry testing, because the concentration of some analytes might vary significantly (i.e., potassium) and, especially, they might produce some degree of hemolysis in the final sample due to excessive aspiration force (e.g., the blood flows through an extremely small opening under a great force). As such, the use of small bore needles should be firmly discouraged in adults with accessible venous accesses, while it might be limited to specific clinical situations, such as patients with very small veins or newborns [68,69]. On the other hand the use of too large needles (e.g., >18-gauge) might also produce hemolysis for the greater likelihood of vein disruption and because the blood might enter the tube faster and more forcefully. Therefore blood aspiration through standard-gauge needles (e.g., between 19- and 21-gauge) is generally advisable; this causes less hemolysis than the use of larger-gauge needles because the flow rate, flow speed, and turbulence are substantially reduced.

6.2.5 Blood collection through spring-loaded and manual lance devices

Skin puncture or capillary blood collection basically involves puncturing the dermis layer of the skin to access the capillary beds that run through the subcutaneous layer of the skin. The blood obtained through this method is a mixture of undetermined proportions of blood from arterioles, venules, capillaries, plus interstitial and intracellular fluids; nevertheless, the proportion of arterial blood is greater than that of venous blood, due to the increased pressure in the arterioles leading into the capillaries versus the pressure in the venules exiting the capillaries.

Blood collection with spring-loaded and manual lance devices is a minimally invasive and easily accessible procedure to obtain capillary blood samples for various laboratory tests in the very young, especially in newborns (e.g., the Clinical Laboratory

Standards Institute [CLSI] recommends capillary blood collection via heelstick for infants less than 1 year of age, whereas capillary blood collection via fingerstick should be considered where appropriate in children older than 1 year) [70]. Capillary blood samples have long been used for obtaining newborn screens and glucose levels. However, thanks to improved laboratory techniques requiring smaller sample volumes and to improved automated heel lancing devices that minimize trauma and pain, this method is viable also in adults to obtain blood for many routine blood tests and preserve venous access for future IV lines. Therefore capillary blood sampling is indicated whenever capillary blood is an acceptable source (e.g., for point-of-care [POC] testing); when the sample volume required is relatively small; in patients with fragile, superficial, or difficult-to-access veins; in cases where multiple unsuccessful venipunctures have already been performed, especially if the test or tests requested require only a small volume of blood; in those with burns or scarring in venous blood collection sites; in extremely obese subjects; and in patients requiring frequent blood tests, receiving IV therapy in both arms or hands, at risk for serious complications associated with venipuncture, venous thrombosis, or deep venous puncture (e.g., deep venous puncture in infants or patients with thrombophlebitis); and in subjects whose veins are "reserved" for IV therapy or chemotherapy. Conversely, capillary blood collection is inappropriate in severely dehydrated patients or those with poor circulation, for coagulation studies requiring plasma specimens and tests requiring large volumes of blood (i.e., erythrocyte sedimentation rate [ESR] and blood cultures). The capillary blood can be used for general clinical chemistry testing, complete blood counts, toxicology, newborn screening, bedside glucose monitoring, and arterial blood gas analysis, it is noteworthy that differences might exist between the reference range of some analytes in capillary blood as compared with venous or arterial blood specimens (e.g., the values of potassium, total protein, and calcium have been reported to be significantly lower in capillary blood, whereas the value of glucose is higher).

Basically two types of lancing devices are used for the collection of capillary blood: puncture devices and incision devices, which are both available in a variety of styles, sizes, and depths. Puncture devices puncture the skin by inserting either a needle or blade vertically into the tissue and are preferred for sites that are repeatedly punctured (e.g., POC blood glucose monitoring or anticoagulation monitoring). On the other hand, incision devices slice through the capillary beds, cause less pain, and require fewer repeat incisions and shorter collection times, so that they are preferable for infant heelsticks.

Several factors should be carefully taken into account before selecting the type of skin puncture device and puncture site, including the patient's age, accessibility of the puncture site, and the volume of blood to be collected. It is also advisable to identify a warm site, which should also be pink and free of any calluses, burns, cuts, scars, bruises, or rashes (e.g., it should not be cyanotic, edematous, or infected). Skin areas that show evidence of previous punctures or are otherwise compromised should also be carefully avoided.

The technique of capillary blood collection is simple, it is occasionally incorrectly performed and may also produce hemolyzed blood as well as other complications and pain for the patient [71]. Meites et al. carried out a systematic study to analyze specimens collected by 15 technologists by using skin-puncture technique involving a Monolet™ lancet. Up to 417 samples were obtained, 176 of which came from newborns and infants up to 13 days of age (one third came from premature infants). Fifty-seven additional samples were obtained from children [72]. The results of this study showed that the mean spuriously free hemoglobin in specimens from pediatric patients was 0.26 g/L, the highest values occurring in newborns 0 to 13 days old (0.39 g/L). Only 3% of the samples exceeded the free hemoglobin value of 1.00 g/L. After 13 days, the technically produced hemolysis was reported to be identical as that for adult plasma (i.e., <0.20 g/L).

Kazmierczak et al. carried out a randomized controlled trial involving 134 newborns assigned to have blood collected; these investigators used either an automated blood collection device (i.e., a spring-loaded device; $n = 68$) or a manual lance ($n = 66$) [73]. A single experienced phlebotomist performed all blood collections. Plasma free hemoglobin concentration was assessed in all specimens to define the extent of hemolysis. No significant differences were observed between the two groups of newborns in terms of gestational age, birth weight, or time interval between birth and time of blood collection, a highly significant difference of plasma hemoglobin concentrations was found between specimens collected with the automated lance (hemoglobin, 2.35 g/L) as compared with those collected using the handheld lance (hemoglobin, 4.85 g/L). Similar results were previously obtained by Paes et al. [74]. These authors investigated 40 healthy, full-term newborns who had blood collected with either the manual lance or a Tenderfoot® device. Analogously, these investigators reported a significant decrease in the concentration of free hemoglobin in specimens collected with the spring-loaded device as compared with the manual lance (0.041 g/L vs. 0.068 g/L, respectively). Overall, the lower incidence of hemolysis associated with use of the automated lance may be explained by the greater blood flow from the cut, the need for less heel manipulation and manual squeezing to obtain the blood sample, as well as the significant increase in the average amount of blood collected with the spring-loaded device versus the manual lance (1.1 vs. 0.6 mL, respectively), and the remarkable reduction in the time needed for specimen collection (120 seconds vs. 212 seconds, respectively). Therefore it might be advisable to collect pediatric or newborn blood through spring-loaded devices, since these reduce the chance that blood specimens will be rejected by the laboratory as being unsuitable for analysis owing to hemolysis and/or an insufficient amount of blood.

The hemolysis occurring during capillary collection has been attributed to a variety of causes, including excessive squeezing, which induces increased hydraulic pressure in the capillaries; the use of a manual lancet, where the depth of the incision is not controlled; the mechanical fragility of the newborns' blood; as well as the newborn's high hematocrit values [72].

6.3 Sample transportation and storage

Appropriate sample transportation is an integral part of obtaining a valid and timely laboratory test result. Because of widespread networking, centralization of laboratory testing within huge facilities, and the need to transport a large number of specimens from peripheral phlebotomy centers to core laboratories, the problem of appropriate conditions of sample transportation (i.e., time, temperature, and humidity) is critically emerging, especially when the primary non-centrifuged specimens are shipped over long distances and/or for excessively long periods of time. In such cases, the death of RBCs causing subsequent hemolysis is typically attributable to the progressive exhaustion of the energy metabolite glucose within the sample. [75]. A significant bias in several analytes (especially LDH, AST, ALT, ALP, potassium, and iron) might, however, also be observed in centrifuged serum tubes without a clot separator owing to prolonged contact of the serum with the clot [76].

In an original study based on the shipment of blood specimens within a commercial transport box from a peripheral collection center to the core laboratory by a land vehicle over a period of nearly 8 hours, it was demonstrated that the recommended temperature for transportation (i.e., between 2°C and 8°C) could be maintained only for a rather limited time (e.g., up to 90 minutes), since the internal temperature was substantially dependent upon the prevailing environmental conditions (i.e., external temperature between 36°C and 38°C) [77]. Most of the problems linking in vitro hemolysis with sample transportation arise from the shipment of whole blood specimens over long distances, since the prolonged contact of serum or plasma with cells can trigger hemolysis. Additional causes include mechanically induced trauma during transport, which can be exacerbated by increased handling when specimens are transported several times (e.g., because of distant collection centers and intermediate sample or transport sorting centers).

The increasing pressures experienced by healthcare professionals to provide more care at less cost and to reduce lengths of hospital stay has translated into a requirement to improve the timeliness of results reporting. Improving the turnaround time (TAT) in the total testing process is a complex task involving the evaluation and monitoring of all the steps from test ordering to results reporting. However, in the hospital setting, the rapid and effective transport of patients' specimens to clinical laboratories represents a key issue, as it strongly affects the total TAT [78]. In fact, delays in TATs are often the result of several processes in pre-, intra-, and postanalytic activities, the method of delivering samples to the laboratory is a prime factor that can have a fundamental impact on both TAT and sample quality [79]. For the purpose of speeding up the specimen delivery process, pneumatic tube systems (PTSs) allow rapid and cost-effective transport of patient specimens to clinical laboratories and therefore are widely used in modern medical centers [80,81]. Depending on the configuration and the speed of the system, a sample transported via PTS may undergo vigorous agitation owing to bends and sudden acceleration and deceleration along its trajectory, resulting in potential modifications of various laboratory parameters.

Although PTSs were introduced in medical centers some decades ago, the evidence of their effects on the quality of patient samples and on all laboratory parameters seems to be limited if not controversial and mostly dependent upon the local environment (i.e., number of angles or turns in the system). In a published paper, hemolysis was measured by visual inspection using a four-point validated scale and defined a priori as greater than 5 g/L of hemoglobin in plasma, so that samples with hemoglobin greater than 5 g/L were rejected. This study contradicted the results previously reported by studies that found a higher hemolysis rate. Studies have evaluated the effects on blood gas analysis of delivering samples to the laboratory through pneumatic tubes [82] as well as effects on coagulation parameters [83], the data regarding hemolysis were obtained some years ago, and as measured by visual inspection or through indirect indexes, such as lactate dehydrogenase and potassium. In addition, in these studies there are few data on the main characteristics of the PTS (namely carrier velocity), and statistical treatment of results does not really allow the defining of the "significance" of observed variations [84,85]. It is also important for laboratory professionals to be aware of the potential drawbacks and defects of PTSs, which may determine rapid sample deceleration and excessive hemolysis. For example, Ellis reported that before the PTS was installed, the weekly mean hemolysis rate was 3.3%, but that this increased to 10.9% after installation. Nevertheless, the burden of hemolyzed specimens increased further to 54% following a system failure. After the fault was corrected and samples were bubble-wrapped, the values fell to 9.0% and to 7.1% [86].

In vitro hemolysis in a blood sample can also occur owing to prolonged storage or storage under inappropriate conditions (e.g., too hot or too cold). In particular, when serum or plasma is in contact with RBCs over a prolonged period, hemolysis may ensue. Placing blood tubes in direct contact with ice or frozen gel packs can also result in hemolysis. Finally, the process of freeze-thawing whole blood should absolutely be avoided.

6.4 Centrifugation of the specimen

Although the length of time between collection and centrifugation is a main determinant of the degree of hemolysis, the extent of remaing tube vacuum during centrifugation does not appear to affect it [48]. As regards centrifugation, mechanical stress inflicted by higher g forces is a well-recognized source of hemolyzed specimens [87]. A mechanical (e.g., gel) barrier not appropriately developed is a further potential source of hemolysis, inasmuch as some RBCs can move back into the serum portion – as, for example, when a sample is respun – since the RBCs will again be in contact with the serum and lysed cells will be homogenized in serum or plasma. Additional sources of hemolysis include centrifugation of the blood before the completion of coagulation and formation of a strong clot and the centrifugation of partially coagulated specimens collected from patients on anticoagulant therapy or samples that have been heparin-contaminated.

7 How to distinguish in vivo versus in vitro hemolysis

One of the most critical issues that the laboratory faces with hemolytic specimens is to discriminate in vitro from in vivo hemolysis, a strategy that should be guided by clinical rather than analytical considerations. At present no consensus exists on how to help clinicians differentiate between the two, and clinical surveillance is highly necessary. The differentiation between these two sources of hemolysis is, however, of paramount importance, as attested by a case study that reported on a patient who died of electromechanical dissociation cardiac arrest due to unnoticed hyperkalemia, as the policy of the laboratory had been not to warn clinicians about potassium results in otherwise hemolyzed samples [88].

Whereas the former circumstance (involving in vitro samples) requires a standardized executive flowchart, encompassing hemolysis detection and quantification, analysis of interference on test results, and recollection of new samples, the latter condition (involving in vivo analysis) may be associated with clinically threatening pathologies that require immediate action. Therefore whenever in vivo hemolysis is suspected, a good liaison between the laboratory personnel and clinical staff is of paramount importance for rapid notification and troubleshooting of the potential underlying cause.

The depletion of plasma haptoglobin is traditionally regarded as a reliable marker for the rapid identification of accelerated in vivo RBC destruction regardless of the site of hemolysis. At variance with other potential markers, haptoglobin levels are not influenced by in vitro hemolysis because the haptoglobin-hemoglobin complexes formed upon RBC breakdown in the circulation are rapidly cleared by monocytes and tissue macrophages via CD163 receptors. This parameter is available for measurement in most clinical chemistry instruments and is therefore suitable for both routine and STAT testing, offering the laboratory a valuable tool whenever in vivo hemolysis must be distinguished from in vitro RBC destruction. It is important to mention here, however, that certain immunological methods might differ in their ability to distinguish hemoglobin-haptoglobin complexes from free haptoglobin. Moreover, since haptoglobin is an acute-phase reactant, the presence of concomitant infections, other reactive states, or chronic hemolysis should be accurately ruled out. A fully mechanized immunonephelometric method for the rapid and specific determination of free hemoglobin in serum, plasma, or aqueous hemolysates was developed more than 20 years ago [89]. The measurement is finalized in nearly 6 minutes, requires 0.25 mL of total sample volume, and the measuring range is rather broad, from 9 to 2,300 g/L of free hemoglobin. Precision is characterized by remarkable intra-assay coefficients of variation (from about 1.6%) and interassay coefficients of variation (from 5.6% to 8.2%). The method is also accurate, yielding a high correlation with hemoglobin determination by cyanohemiglobin spectrometry ($r = 0.984$). Interference has, however, been reported for lipemic samples. An additional technique for the assessment of free and protein-bound hemoglobin in plasma using a combination of high-pressure liquid chromatography (HPLC) and absorption spectrometry has been developed, but this assay is obviously unsuitable for systematic and routine screening for hemolyzed plasma or serum specimens [90].

Rather, it might be used as an alternative to the traditional cyanohemiglobin spectrometry assay to set target values for this analyte within external quality assessment schemes. It is also worth mentioning that a strategy based on either the immunonephelometric or HPLC assays currently seems unsuitable since it would increase the test TAT, since these assays are mostly unsuitable for STAT testing instruments, are relatively expensive, and have already been surpassed by the hemolysis index (HI, Chapter 10).

A new flow cytometric method has also been developed to detect damaged RBCs using antihemoglobin antibodies in hypotonic solution. Although the performance of this test is indeed optimal for clinical use, its use in daily laboratory practice to screen for hemolyzed specimens is unsuitable for obvious technical, economic, and practical reasons [91].

Reticulocytosis is frequent in hemolytic anemia but is poorly specific, since increased RBC production is also associated with blood loss or bone marrow response to iron, vitamin B12, and folate deficiencies. The reticulocyte count can also be close to normal in patients with bone marrow suppression despite ongoing severe hemolysis. High MCH and MCHC values are also suggestive of spherocytosis. Serum LDH elevation is commonplace in hemolysis, but it lacks diagnostic specificity because the enzyme is ubiquitous and is also released from a variety of neoplastic or ischemic cells elsewhere in the body. LDH isoenzymes 1 and 2 are more specific for RBC destruction, but they are also released from the injured myocardium. Increased concentration of unconjugated bilirubin is observed in patients with in vivo hemolysis, but it is also a hallmark of Gilbert's disease. In the presence of hemolytic anemia, the concentration of indirect bilirubin is usually <4 mg/dL, so that higher values mirror the presence of impaired liver function. The presence of free hemoglobin in urine reflects hemoglobinuria, occurring with intravascular hemolysis when the amount of free hemoglobin exceeds the concentration of available haptoglobin in the bloodstream. The urine might be dark in the presence of hemoglobinuria, but myoglobinuria, porphyria, and other conditions also lead to dark urines. Taken together, increased values of LDH and decreased haptoglobin rather than an augmented potassium concentration are the most sensitive general tests for diagnosing hemolysis and distinguishing between spurious hemolysis and hemolytic anemia. In fact, the release of potassium during intravascular hemolysis it typically distributed throughout all body fluids, which would reduce the otherwise increased concentration arising from in vivo hemolysis. The potassium excess is also effectively cleared by the kidneys even if the equilibrium in the plasma is reached within 1 hour, so that only 5% to 10% of the potassium load can be eliminated early by the kidneys. Conversely, LDH is a larger molecule (i.e., 160 kDa), which remains in the vascular space much longer; it is therefore characterized by a slower clearance [92].

When in vivo hemolysis is suspected, some additional simple tests in combination with the clinical history, physical examination, peripheral blood smear, and laboratory investigations previously described are necessary for the differential diagnosis. A complete blood count will help to diagnose anemia as well as leukocyte abnormalities, whereas a platelet count rules out an underlying infection or hematologic malignancy, since its value is within the reference range in most hemolytic anemias. Thrombocytopenia can occur in systemic lupus erythematosus, chronic lymphocytic leukemia (CLL), and microangiopathic hemolytic anemia. Peripheral smear and morphologic examination can reveal the presence of polychromasia (reflecting RBC immaturity), reticulocytosis, and spherocytes (e.g., suggestive of congenital spherocytosis), and schistocytes, but it can also help diagnose a concomitant underlying hematologic malignancy associated with hemolysis (e.g., CLL).

RBC indices are also useful. Low levels of MCV and MCH are consistent with a microcytic hypochromic anemia, which may occur in chronic intravascular hemolysis. A high MCV is consistent with a macrocytic anemia, usually due to megaloblastic anemias or liver disease. The direct Coombs test is usually positive in patients with AIHA, but it may occasionally be negative in up to 10% of cases. A high titer of anti–I antibody might be found in mycoplasmal infections and a high titer of anti–I antibody may be observed in hemolysis associated with infectious mononucleosis. An anti–P cold agglutinin can be detected in paroxysmal cold hemoglobinuria. A G6PD screen is useful to diagnose a deficiency of this enzyme, but results can be normal when the reticulocyte count is elevated because immature RBCs contain a considerable amount of G6PD. A Heinz body preparation can also help to detect G6PD deficiency.

8 Effects of in vitro hemolysis on laboratory testing

In vitro hemolysis traditionally reflects a more generalized process of vascular and blood cell damage that has occurred during phlebotomy, causing cell membrane disruption and leakage of hemoglobin and other cellular components into the surrounding fluid. Therefore the analysis of hemolyzed specimens can produce important (and unwanted) effects on laboratory testing and lead to inaccurate results.

The International Federation of Clinical Chemistry and Laboratory Medicine (IFCC) provides a clear definition of analytical interference, which is "the systematic error of measurement caused by a sample component, which does not, by itself, produce a signal in the measuring system" [93]. However, Selby has provided another suitable definition, which is "the effect of a substance present in an analytical system, which causes deviation of the measured value from the true value" [94]. Therefore hemolysis is inherently a challenging problem in laboratory testing, since it may not be evident until whole blood specimen centrifugation has been performed, exposing the serum or plasma to specific scrutiny. Moreover, hemolysis can significantly influence the reliability of a variety of tests for a multitude of biological and analytical reasons. One of the first studies to unmask the interference of in vitro hemolysis in laboratory testing was published more than 60 years ago by Sobel and Snow, who showed that hemolyzed blood has been found to give higher carotene values, and because of the high blank values after irradiation, made the final results unreliable [95]. In a further investigation Utley et al., using the micromethod of Bessey and Lowry, also found that in vitro hemolysis caused a false increase in serum vitamin A values. They concluded that serum samples showing evidence of more than a trace of hemolysis should be discarded [96].

Although the first question to be answered is indeed whether laboratory professionals are aware of the problem that some laboratory tests might be affected by hemolysis, the "European Preanalytical Scientific Committee (EPSC) – International Federation of Clinical Chemistry and Laboratory Medicine (IFCC) Working Group on Laboratory Errors and Patient Safety (WG-LEPS)" survey highlighted that 91% of participants do indeed acknowledge this problem (▶Tab. 4.2). However another survey, designed to assess knowledge of the expected elevation in serum potassium measurement with different degrees of hemolysis, administered a questionnaire specifically to medical technologists working in biochemistry laboratories, hospital physicians, and nurses. As a result, it was found that healthcare staff tended to overestimate the interference and therefore used an incorrect proportional adjustment approach to the problem. It was hence concluded that this poor knowledge and faulty thinking could lead to diagnostic delays or misdiagnoses [97].

8.1 Interference in clinical chemistry testing

Several studies have investigated the influence of hemolysis on clinical chemistry testing by means of different techniques for obtaining artifactual hemolysis and using different instrumentations, test principles, and reagents.

More than 30 years ago, Laessig et al. carried out one of the first systematic studies to evaluate the effects of 0.1% and 1.0% RBCs and hemolysis on serum chemistry values [76]. Basically, two degrees of interference, corresponding to removal of 99% and 99.9% of the erythrocytes, were used to examine the effects of both hemolyzed and intact cells on 40 clinical chemistry assays. Serum lactate dehydrogenase (LDH) activities were those most strongly affected by hemolyzed erythrocytes, whereas potassium, creatine kinase (CK), aspartate aminotransferase (AST), alanine aminotransferase (ALT), and iron showed smaller but significant effects because the samples contained 1% of hemolyzed cells, with minor bias being recorded at the 0.1% level. The presence of non-hemolyzed cells at either level did not significantly modify the results of the remaining clinical chemistry parameters.

A few years later, Frank et al. also assessed the effect of in vitro hemolysis on clinical chemistry values in serum [98]. Erythrocyte hemolysates were prepared from EDTA-anticoagulated whole blood samples. The RBCs, separated by centrifugation, were washed three times with isotonic saline and then lysed by adding an equal volume of deionized water and sodium saponin. The hemoglobin concentration was assessed by the benzidine method. Sufficient hemolysate was then spiked to serum samples to obtain a wide range of free hemoglobin concentrations between 0.09 and 2.8 g/L (the latter hemoglobin concentration gave a bright red hue to the specimen, consistent with gross hemolysis). As a result, potassium values were the most biased, showing a linearly direct increase as a function of the free hemoglobin content of the serum specimen (i.e., roughly 0.75 mmol/L for hemoglobin concentrations from 0.09 to 2.8 g/L). A significant decrease was also observed for bicarbonate, whereas serum iron variably decreased, depending on the assay used. As regards serum enzymes, LDH appeared to be most sensitive to hemolysis (i.e., the increase of 260 U/L of LDH resulted in a 2.5-fold higher enzyme activity at 2.8 g/L of free hemoglobin in serum as compared with the serum specimen containing 0.09 g/L of free hemoglobin). Both CK and acid phosphatase showed a specific methodological bias. Therefore acid phosphatase measured with prostatic isoenzyme–specific substrates exhibited a negligible increase in enzyme activity, whereas the increase in total acid phosphatase activity was 4-fold. The increase in the enzyme activity of CK spanned from a minimum of 1.4-fold to a maximum of 2.3-fold, depending on the instrumentation used. A remarkable decrease of bilirubin concentration was also observed in hemolyzed serum specimens.

In a further investigation, Blank et al. added increasing concentrations of purified human hemoglobin (human isoionic oxyhemoglobin), up to 19.3 g/L, to various serum pools to verify whether the presence of free hemoglobin in the sample might interfere with the determination of a variety of clinical chemistry analytes [99]. It was hence shown that the hemolysate generated a positive bias in the measurement of direct bilirubin, total protein, albumin, and AST, whereas results for creatinine were negatively biased with one instrument (i.e., Astra) but not with another (i.e., SMAC).

Sonntag assessed the interference of hemolysate in the determination of 26 clinical chemical parameters using a variety of fully mechanized analytical instruments [100]. Hemolysis was obtained by storage of whole anticoagulated blood for 20 minutes at −40°C in a deep freezer. Increasing amounts of hemolysate were then spiked to non-hemolytic serum in order to simulate the conditions in practice. The results of this experiment showed that hemolysis interferes in the determination of the concentrations or catalytic concentrations of ALT, albumin (by electrophoresis), alkaline phosphatase (ALP), AST, bilirubin, CK, α2-globulin (by electrophoresis), β-globulin (by electrophoresis), GGT, potassium, LDH, and acid phosphatase. The values of albumin, ALP, and α2-globulin were decreased, whereas those for all the other analytes listed in the previous sentence were increased. Moreover, hemolysate at final hemoglobin concentrations up to 6.6 g/L caused no clinically relevant interference in the determination of albumin (immunonephelometric), α-amylase, calcium, cholinesterase, chloride, cholesterol, creatinine, iron, glutamate dehydrogenase, α1-globulin (by electrophoresis), γ-globulin (by electrophoresis), glucose, uric acid, urea, sodium, phosphate, transferrin, and triglycerides.

Randall et al. – as part of an evaluation of a Synchron CX5 analyzer – carried out an original investigation to define the interference of hemolysis, bilirubin, and lipemia on a wide range of clinical chemistry tests, including urea, creatinine, uric acid, total protein, albumin, calcium, total bilirubin, ALP, AST, gamma-glutamyl transferase (GGT), and inorganic phosphate. Two types of interferences were observed, the former also described in other analyzers and attributed to analytical problems with the measurement of a particular analyte, the latter as a consequence of the bichromatic optical system used on the CX-5 (e.g., interference of free hemoglobin in the measurement of total protein and inorganic phosphate) [101].

Yücel and Dalva also examined the effects of hemolysis on the results of 25 common biochemical tests by collecting 60 blood samples from inpatients and outpatients. Part of the whole blood specimens was mechanically hemolyzed in a two-step procedure (i.e., by stirring the sample with a metallic bar for 1 and 5 minutes to obtained moderate and severe hemolysis). The free hemoglobin in all specimens was then measured spectrophotometrically and the samples were grouped as being non-hemolyzed, moderately hemolyzed, and severely hemolyzed [102]. Further clinical chemistry testing on the samples revealed that hemolyzed specimens had decreased values of glucose, uric acid, ALP, lipase, and total and direct bilirubin as well as slightly increased values for AST, ALT, high-density lipoprotein (HDL) cholesterol, triglycerides, and inorganic phosphate, although these changes did not reach statistical significance even in the severely hemolyzed group. The values of total protein, albumin, urea, creatinine, sodium, chloride, calcium, magnesium, and amylase were unchanged when hemolyzed specimens were compared with reference non-hemolyzed specimens. Therefore only the values of LDH, potassium, acid phosphatase, and prostatic acid phosphatase were significantly increased in moderately hemolyzed specimens, and the increase in concentration of these parameters was even more significant in severely hemolyzed samples, along with that of total cholesterol.

Jay and Provasek not only characterized hemolysis interference but also tested a mathematical correction for selected Hitachi 717 assays [103]. Separated RBCs were lysed with deionized water and frozen (1 hour at –30°C), and then spiked to serum pools in order to achieve final free hemoglobin concentrations of approximately 0.5, 1.0, 2.0, 4.0, and 5.0 g/L. Error was calculated in both relative and absolute terms and was plotted as a function of hemoglobin concentration. Clinical significance was judged relevant when absolute error exceeded 10% of the result obtained on non-hemolyzed specimens. Correction factors were calculated by regression analysis of absolute error on hemoglobin concentration. A clinically significant negative bias was observed for ALP, α-amylase, and GGT, whereas a significant positive interference was found for AST, CK, LDH, and potassium. Variable interference was observed for bilirubin and theophylline. An absolute error could be calculated by linear regression analysis on the hemoglobin concentration for all the analytes tested except bilirubin and theophylline. Moreover, the amount of error could be reliably predicted by multiplying the hemoglobin concentration by the slope obtained from linear regression, even though the authors raised concern about the practical use of these corrective formulas, inasmuch as interindividual differences exist in intracellular analyte concentrations, which could produce variability in correction factor determination.

Grafmeyer et al. assessed the influence of bilirubin, hemolysis, and turbidity on 20 analytical tests performed on automatic analyzers [104]. The analyses entailed 13 substrates/chemistries including albumin, calcium, cholesterol, creatinine, glucose, iron, magnesium, phosphorus, total bilirubin, total proteins, triglycerides, uric acid, and urea as well as the enzymatic activities of ALP, ALT, alpha-amylase, AST, CK, GGT, and LDH, which were assessed on 15 automatic analyzers (Astra 8, AU 510, AU 5010, AU 5000, Chem 1, CX 7, Dax 72, Dimension, Ektachem, Hitachi 717, Hitachi 737, Hitachi 747, Monarch, Open 30, Paramax, Wako 30 R). Using a solution of hemolyzed erythrocytes in distilled water, plasma pools were prepared containing hemoglobin concentrations in the range of 0 to 3.86 g/L. It was thereby observed that hemolysis interfered in 120 of the 348 tests (34.5%) performed for the 20 parameters in the 15 analyzers.

Hübl et al. compared the enzymatic determination of sodium, potassium, and chloride in abnormal (i.e., hemolyzed, icteric, lipemic, paraproteinemic, or uremic) serum samples with indirect determination by ion-selective electrodes [105]. In particular, 33 hemolyzed serum samples (mean ± standard deviation serum hemoglobin 2.83 ± 1.90 g/L, range 1.19 to 8.99 g/L), were assayed in parallel on the standard ISE unit of a Hitachi 717, 737, or 911 and the corresponding enzymatic procedures on the same instrumentation (e.g., sodium and potassium analyses were performed by determining Na-dependent β–galactosidase and K-dependent pyruvate kinase activities respectively, whereas chloride was enzymatically determined by measuring chloride-dependent α-amylase activity). The enzymatic results were in good agreement with those by ISE, the interference-related differences generally being without clinical significance.

Lippi et al. spiked serum samples with hemolysate obtained after freeze-thawing whole blood specimens and measured several clinical chemistry parameters on the Roche Modular P, demonstrating that the interference appears to be approximately linearly dependent on the final concentration of blood cells lysate [106]. Therefore the addition of blood hemolysate generated a consistent and dose-dependent trend toward overestimation of ALT, AST, creatinine, CK, iron, LDH, lipase, magnesium, phosphorus, potassium, and urea, whereas mean values of albumin, ALP, bilirubin, chloride, GGT, glucose, and sodium were substantially decreased compared with the baseline specimens containing no hemolysate. Statistically significant differences in samples containing up to 0.6 g/L of free hemoglobin in serum or less could already be observed for AST, LDH, potassium, and sodium. Albumin, total bilirubin, calcium, chloride, and uric acid measurements were not significantly affected, even in the presence of up to 20.6 g/L of serum hemoglobin. Clinically meaningful variations, as reflected by deviation from the current analytical quality specifications for desirable bias, were observed for AST, chloride, LDH, potassium, and sodium in samples with mild or almost undetectable hemolysis by visual inspection (i.e., 0.6 g/L of free hemoglobin in serum). Among all the analytes tested, the bias values recorded for total bilirubin, calcium, and glucose were the only ones that did not achieve a clinically significant variation, even at the highest concentration of serum hemoglobin (i.e., 20.6 g/L). Despite the approximately linear relationship of variations to free hemoglobin concentration in serum, an unpredictable, heterogeneous, and sample-specific response was observed for several parameters.

Steen et al. promoted a multicenter evaluation of the interference of hemoglobin as well as bilirubin and lipids on Synchron LX-20 assays [107]. Hemoglobin was added to serum specimens in the form of an erythrocyte hemolysate prepared by osmotic disruption with distilled water. Different dilutions were prepared starting from pooled patient serum samples without any detectable interference at a normal and, as far as possible, an abnormal analyte concentration. Analytically relevant interference by hemoglobin (the cutoff for free hemoglobin interference is reported within parentheses) was observed for amylase and AST (at 0.3 g/L); LDH and total bilirubin (at 0.8 g/L); potassium, magnesium, GGT, uric acid, and ammonia (at 1.3 g/L); and chloride and creatinine (at 3.2 g/L). The cutoff indices proposed by the authors showed a very good correlation with those recommended by the manufacturer.

Koseoglu et al. enrolled 16 healthy volunteers and generated four hemolysis levels by mechanical trauma according to hemoglobin concentrations as follows: 0 to 0.10 g/L (group I), 0.10 to 0.50 g/L (group II), 0.51 to 1.00 g/L (group III), 1.01 to 2.50 g/L (group VI) and 2.51 to 4.50 g/L (group V) [108]. The presence of hemolysate significantly affected LDH and AST at almost undetectable hemolysis by visual inspection (i.e., free hemoglobin in plasma <0.5 g/L). Moreover, clinically meaningful bias in potassium and total bilirubin were recorded on moderately hemolyzed samples (free hemoglobin in plasma > 1 g/L). Interference in ALT, cholesterol, GGT, and inorganic phosphate concentrations were instead not observed up to severely hemolyzed levels (free hemoglobin in plasma 2.5 to 4.5 g/L). Albumin, ALP, amylase, chloride, HDL cholesterol, CK, glucose, magnesium, total protein, triglycerides, unsaturated iron binding capacity, and uric acid also exhibited statistically significant variations, which, however, always remained within the limits of the Clinical Laboratory Improvement Amendments (CLIA).

8.1.1 Mechanism of interference

Basically the interference in hemolyzed samples is due to the leakage of hemoglobin as well as other intracellular components into the surrounding fluid. This phenomenon induces false elevations of some analytes, dilutional effects on others, chemical interference of free hemoglobin in several analytic reactions, as well as method – and analyte concentration – dependent spectrophotometric interference. The spectrophotometric interference is mainly attributable to increased optical absorbance or change in the blank value and is particularly higher for those laboratory tests that require measurements at 415, 540, and 570 nm since at these wavelengths the free hemoglobin molecules more strongly absorbs (▶Fig. 8.1) [5,104,106,109,110]. Clearly, at high levels of serum hemoglobin, these interferences might coexist, producing biases that do not necessarily proceed in the same direction and thus making the outcome (i.e., both the direction and the extent of the bias) often unpredictable.

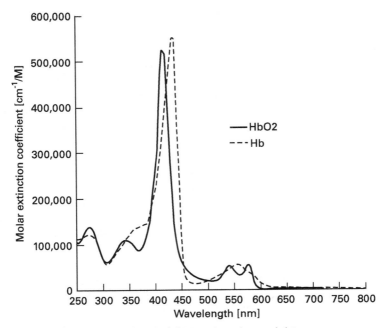

Fig. 8.1: Peak of absorbance of hemoglobin and oxyhemoglobin.

8.1.2 Interference due to the delta between intracellular and extracellular concentrations of the analytes

Albumin, ALP, bilirubin, chloride, GGT, glucose, and sodium are characterized by a higher extracellular than intracellular concentration and might thus, because of dilutional effects, be significantly reduced in hemolyzed specimens. Conversely, the observed increase in AST, LDH, magnesium, phosphorus, and potassium levels in hemolyzed specimens is attributed to greater intracellular concentrations. In particular, the erythrocytes contain about 160-fold as much LDH, some 68-fold as much acid phosphatase, about 40-fold as much AST, and about 6.7-fold as much ALT as does plasma [106, 111].

It is, however, noteworthy that although the intracellular concentration of magnesium is nearly three times greater than its plasmatic values, the extent of hemolysis must be very large to produce a clinically significant effect [104]. The RBCs also contain a large concentration of organic phosphate esters, and the concentration of inorganic phosphate in hemolyzed samples is thereby increased because these esters are hydrolyzed by serum phosphatases. The case of the interference of hemolysis in iron testing is even more interesting and challenging. One μmol of hemoglobin monomer contains one μmol of iron. The presence of 3.86 g/L of free hemoglobin in the specimen would thereby produce a dramatic increase in iron test results; but, for unknown reasons, certain iron assays are less sensitive than others to such bias, even on occasion showing a negative bias. Grafmeyer has suggested that this might be dependent on the pH of the reaction medium, the detergents used, and spectral interference. In particular, when iron is not released from hemoglobin molecules, the interference is probably optical in nature and might be eliminated using a sample blank. Conversely, when iron is released from hemoglobin molecules, a substantial interference might be commonplace [104]. Because the hemoglobin is a protein, its presence in free form in hemolyzed samples would also contextually increase the results of total protein testing. Despite the fact that even in the presence of massive hemolysis (i.e., up to 3.86 g/L of free hemoglobin in the specimen) the relative amount of intracellular proteins released would be modest, this amount would be still meaningful in relative terms (e.g., about 3.8 g/L for 70 g/L and about 3 g/L for 65 g/L). The bias might even be greater because of additional spectral interference.

8.1.3 Spectrophotometric interference

Hemoglobin strongly absorbs at 415 nm (the "Soret peak") and to a lesser extent between 530 and 600 nm, having two peaks at 540 and 570 nm. Therefore it may greatly interfere with enzymatic assays encompassing measurement at these wavelengths. The structure of hemoglobin might vary with the medium, precipitating in acid media, converting to hematin in alkaline media, becoming reduced, forming methemoglobin or even carboxy or sulfohemoglobin, all of which exhibits different spectra and diverse patterns of interference [104]. Nevertheless, the spectrophotometric interference is roughly linear and dependent upon the final concentration of free hemoglobin in the serum or plasma, which might generate a consistent trend towards the overestimation of some enzymes such as AST, ALT, CK, and LDH. The effects of such interference on bilirubin, iron, lipase, and GGT measurement is mostly attributable to spectral overlap in a chemical reaction between free hemoglobin and reagents, whereas that on ALP has been ascribed to the denaturation of hemoglobin during reaction in an alkaline medium [107].

8.1.4 Chemical interference

The presence of several "abnormal" substances in a reagent solution might generate interference at different stages of the reaction, consuming some of the products. The RBCs contain several structural proteins, enzymes, lipids, and carbohydrates besides hemoglobin, and many of these may interact or compete with the assay reagents [5]. The effect of hemoglobin on bilirubin measurement is a paradigmatic example, since azobilirubin is measured at 540 nm, corresponding to one of the wavelengths at which oxyhemoglobin absorbs. Increased concentrations of CK (which occur in hemolyzed specimens) might also be explained by a well-known chemical interference sustained by the release of intracellular adenylate kinase, which is not completely inhibited under reaction conditions. When the pH varies during the reaction, hemoglobin might be transformed into hematin or methemoglobin and, contextually, the bilirubin being measured is destroyed by hemoglobin peroxidase. Therefore, using the Jendrassik-Gróf method, falsely low concentrations of bilirubin are typically observed because the pseudoperoxidase activity of hemoglobin inhibits the formation of the azo dye (the inhibition is typically observed once the free hemoglobin concentration in serum or plasma is as high as 0.8 g/L). A positive bias at normal neonatal bilirubin concentrations and a negative one at intermediate and high neonatal bilirubin concentrations has also been reported in capillary heelstick samples [112]. Only high cell-free hemoglobin concentrations cause lower serum values of uric acid; the uricase-catalase method (Kageyama reaction) is, however, more susceptible to interference than the uricase-peroxidase method [5] (▶Tab. 8.1).

Tab. 8.1: Leading causes of hemolysis interference in clinical chemistry testing

1. False elevations of analytes due to leakage of intracellular components into the surrounding fluid	
a. Hemoglobin	↑↓
b. Potassium	↑
c. Lactate dehydrogenase (LDH)	↑
d. Aspartate aminotransferase (AST)	↑
e. Alanine aminotransferase (ALT)	↑
f. Creatinine	↑
g. Creatine kinase (CK)	↑
h. Iron	↑
2. Dilutional effects	
a. Albumin	↓
b. Chloride	↓
c. Glucose	↓
d. Sodium	↓
3. Chemical and spectrophotometric interferences	
a. Alkaline phosphatase (ALP)	↓
b. Bilirubin	↓/↑
c. Creatine kinase (CK)	↑
d. Iron	↑
e. γ-glutamyl-transferase (GGT)	↓
f. Lipase	↓
g. Magnesium	↓
h. Phosphorus	↓
i. Urea	↓

8.1.5 Additional considerations regarding interference

It is also important to mention that a different test is not affected to the same extent by the degree of hemolysis. This is also true in using different assays for measuring the same analyte, even on the same analytical platform. Gobert De Paepe et al. showed that one method (Bilirubin 2, Siemens) for the quantitative determination of total and conjugated bilirubin was less sensitive to hemolysis than two other assays (Synermed, Sofibel; Bilirubin Auto FS, Diasys), even at low bilirubin levels. The measurement of conjugated bilirubin was also significantly altered by low hemoglobin concentrations for both Bilirubin Auto FS and Synermed. In marked contrast, the authors observed no hemoglobin interference with the Direct Bilirubin 2 reagent, which complied with the method validation criteria from the French Society for Biological Chemistry [113].

Dimeski et al. investigated whether a relationship exists between an increase in total lipid concentration (cholesterol plus triglycerides) and hemolyzed specimens, by analyzing all samples collected for lipid analysis over a two-year period [114]. The degree of hemolysis was measured as HI on a Hitachi Modular System with Roche reagents. A multiple linear regression analysis of ln(HI) values indicated that both cholesterol and triglycerides were linearly (ln) related to the HI. The data at higher lipid concentrations also suggested that the triglyceride concentration had a greater influence on the HI than the cholesterol concentration, thereby demonstrating that increasing hyperlipidemia (particularly hypertriglyceridemia) is associated with increased hemolysis. Whether this association is causal or causative, however, is yet to be established.

It is also important to emphasize that it is strictly necessary to evaluate the influence of hemolysis on all tests and to state the limits of acceptability for each as proposed in validation protocols [115,116] and as discussed elsewhere in this book. Therefore maximal tolerable systematic error and inaccuracy should be provided for biological specimens and compared with those obtained using a reference or validated method.

8.2 Interference in hemostasis testing

The complexity of hemostasis along with the sophistication of modern coagulation analytical techniques makes the coagulation test among the most prone to preanalytical errors in the clinical laboratory. Accordingly, the quality of the blood specimen is of foremost importance to most hemostasis tests, especially those based on clotting assays [44,117,118]. To some extent, whether hemolysis affects test results depends on the instrument used to perform testing. Current instruments used in hemostasis testing may detect assay end points in various ways, including (a) mechanical (detects physical formation of fibrin strands), (b) viscosity-based (viscosity of plasma increases as clot formation begins) (c) photo-optic (e.g., an increase in turbidity is noted with fibrin formation; a change in absorbance is detected in some chromogenic assays), and (d) centrifugal nephelometry, in which increasing fibrin formation gives rise to an increase in light scatter. Accordingly, the effect of hemolysis as related to colorimetric interference is minimal with mechanical test methods.

Nevertheless, interference in hemostasis testing involves more than just colorimetric interference, with the release of cell lysates, including adenosine diphosphate (ADP), that can influence platelet activity, and with cell-derived tissue factor and phospholipids activating various coagulation pathways [69,119]. In brief, the potential interference from in vitro hemolysis is thus not entirely due to the presence of free hemoglobin in plasma because many other substances involved in activating coagulation can be released from blood cells. Moreover, the meaning of "hemoglobin in the sample" is thereby not equivalent to "hemolysate," at least for coagulation testing, since the presence of debris, protein, and other potential interferents from cells additionally contributes to the effects of the hemolysate. Therefore it is very likely that the source of bias in hemostasis assays would arise mainly from the release of intracellular and thromboplastic substances from both leukocytes and platelets.

Thus the Clinical and Laboratory Standards Institute, in its guidelines for prothrombin time (PT) and activated partial thromboplastin time (APTT) testing, recommends that samples with visible hemolysis should not be used because of possible clotting factor activation and interference with endpoint measurement [120].

Lippi et al. have reported prolongations in prothrombin time (PT) and increased levels of D-dimer in samples containing final lysate concentrations of 0.5% and 2.7%, respectively. In samples containing a final lysate concentration of 0.9%, significant shortenings of activated partial thromboplastin time (APTT) and decreased values of fibrinogen have instead been observed [69]. To determine whether hemolysis might influence the reliability of PT and APTT test results, Laga et al. carried out a prospective two-phase study [119]. In the former phase, it was investigated whether a difference in PT, APTT, and selected factor assays can arise between hemolyzed and subsequently recollected non-hemolyzed specimens from the same patient. More reliably, in the latter phase, the samples were obtained from healthy human subjects, and in vitro hemolysis was induced experimentally by mechanical means (i.e., by using a tissue homogenizer) to determine whether a relationship exists between the degree of hemolysis and any observed change in test results. The supernatant hemoglobin values ranged from 0.2 to 7.0 g/L in the hemolyzed specimens and 0 to 0.3 g/L in the non-hemolyzed specimens. A nonsignificant shortening of PT (15.8 ± 8.4 vs. 16.3 ± 8.7 seconds) and a significant prolongation of APTT (31.6 ± 18 vs. 32.5 ± 19, p <0.01) were observed in hemolyzed as compared with non-hemolyzed specimens. Although statistically significant, the overall difference was not considered clinically meaningful. Differences were also observed for assays of clotting factors of the extrinsic pathway (factor VIIa, V, X) but not for those of the intrinsic pathway (factor XIIa, factor VIII). At variance with the previous study, which used a different method of lysis (freeze-thaw), a very modest and nonsignificant trend for change was apparent in the PT despite the progressive increase of free hemoglobin in the supernatant, whereas a progressive, statistically but not clinically significant lengthening of the APTT was recorded (repeated measures ANOVA F = 5.2; p <0.01). It is, however, noteworthy that the authors concluded that the hemolyzed specimens they tested, which reliably mirrored those typically obtained in an outpatient setting, might be suitable for the analysis of PT, neither hemolyzed nor non-hemolyzed specimens were suitable for delayed (>3 hours) APTT testing, which is in agreement with the current guidelines. A significant decrease of antithrombin (assayed using a chromogenic method) was also found in hemolyzed specimens [119] (▶Tab. 8.2).

In a recent article it was also shown that hemolysis substantially modified thromboelastography (TEG) variables and platelet activation indices. Basically hemolysis determined a reduction of reaction time and alpha angle while it simultaneously increased the coagulation time. Hemolysis also resulted in decreased mean platelet component (MPC) concentration [121]. Importantly, although the degree of hemolysis can be assessed in plasma samples destined for routine coagulation based tests using plasma or more specialized assays such as antithrombin, inspection is not possible in whole blood samples as used for TEG and blood indices. Studies related to hemolysis and platelet function testing using classic aggregation analysis are essentially lacking [122], it would be expected that cellular releasates would similarly adversely affect

Tab. 8.2: Leading causes of hemolysis interference in coagulation testing

Prothrombin time (PT)	↑
D-dimer	↑
Activate partial thromboplastin time (APTT)	↓
Antithrombin	↓

test parameters and give rise to clinically significant differences. The effect would possibly be more problematic using whole blood aggregometry [123] than light transmittance aggregometry [124], since visual inspection for hemolysis, like TEG, is not easily applied for the former. Similarly, the effect of hemolysis on the PFA-100 closure time, a popular platelet function screening test [125], would also be expected to generate a clinically significant effect, but with this again not being easily assessable on a case-by-case basis.

Gottschall et al. added hemolyzed RBCs in the presence of dilute plasma in quantities equivalent to as little as 0.2% hemolysis and observed a significant elevation in the apparent value of IgG associated with platelets (PAIgG). The addition of hemolysate to plasma containing low concentrations of platelets, which thereby simulates that obtained from thrombocytopenic patients, caused a more striking elevation in PAIgG. Further testing using simulated patient whole blood samples anticoagulated with EDTA yielded identical results. It was therefore concluded that caution should be exercised in interpreting the results of PAIgG measurements on platelets isolated from blood samples in which even minimal in vitro hemolysis has occurred [126].

8.3 Interference in immunoassays and electrophoresis

Regardless of the efforts of developers and manufacturers of clinical laboratory tests and the vigilance of laboratory staff, some tests are occasionally subjected to interferences that can affect the reliability of results. Immunoassays are no exception to this rule, and the problem might even be magnified since sample interferents such as hemolysis, icterus, and lipemia cannot be measured reliably on most immunoassay instruments. A huge body of evidence has been gathered to demonstrate either false-positive or false-negative results in immunoassays or else reports investigations of specific interferences, which include plasma, serum proteins (e.g., rheumatoid factor, binding proteins), heterophile and antianimal antibodies, drugs and drug metabolites, agar, and cross-reacting substances as well as hemolysis [127,128].

The interference from hemolysis in immunoassays is probably less important than in spectrophotometric methods, a more generalized problem of cellular debris resulting from blood cells lysis, rather than simply the presence of free hemoglobin in plasma or serum, may (as in hemostasis testing) have a significant effect on some analytes and methods [129]. The potential mechanisms include interference with antigen-antibody reaction, with color measurement in assays, and with the enzyme reactions in enzyme-labeled antibody detection assays, which affect homogenous enzyme-antibody assays more than heterogeneous assays. Qualitative antibody assays that are heterogeneous with a washing step do not show much interference with hemolysis. Interference can also occur more frequently when antibodies used in immunoassays cross-react with some compounds released from the lysed blood cells or some of the released compounds bind to the analyte and thereby mask the antibody binding sites. Hemolysis can also produce interference by accelerating or suppressing immunoreactions [130].

So far interference has been detected in several specific circumstances. In particular, the influence of hemolysis on cardiac troponin testing has been deeply reviewed [131]. Lyon et al. as well as Sodi et al. showed falsely decreased levels of troponin T due to hemoglobin per se and/or proteolysis [132,133], whereas Masimasi and Means reported on elevated troponin I and prostate-specific antigen (PSA) levels [134]. Li and Brattsand spiked hemolysate obtained from freezing EDTA blood to two serum samples and also observed a negative bias on troponin T in a cell-free hemoglobin concentration-dependent fashion [135]. Dasgupta et al. also investigated the effect of hemolysis on two troponin I immunoassays from Abbott Diagnostics (AxSYM TnI microparticle enzyme immunoassay, MEIA) and Bayer Diagnostics (ACS:180 TnI chemiluminescent immunoassay, CLIA) [136]. Small amounts of a hemoglobin solution were added to two serum pools and a modest and non–statistically significant decrease in troponin I was observed with moderate concentrations of hemoglobin, up to 0.2 g/L, on AxSYM TnI. Both immunoassays were, however, negatively biased (up to −40%), with cell-free hemoglobin concentrations above 0.4 g/L. Hawkins studied the effect of hemolysis on the Ortho-Clinical Diagnostics Vitros ECi and Beckman Coulter Access 2 AccuTnI [137]. The hemolysate, prepared from the lysis of packed erythrocytes with water and freezing, were added to non-hemolyzed plasma pools. No false-positive cases occurred with the Access TnI for any concentration of troponin I, but the false-positive rate increased with the ECi TnI immunoassay from values of 0.22 µg/L of troponin I Lippi et al. hemolyzed 12 K2EDTA-anticoagulated samples by one- or two-time aspiration through a thin needle (30-gauge) and tested them with the modified Beckman Coulter Accu-TnI immunoassay. The concentration of troponin I was progressively decreased in hemolyzed specimens, exceeding a 20% variation (i.e., the significant variation currently proposed by the National Academy of Clinical Biochemistry) in only 3 out of the 12 cases containing cell-free hemoglobin >16.5 g/L [131]. Daves et al. [138] prepared serial aliquots of homologous hemolyzed samples, collected from 14 individuals, containing final concentrations of plasma hemoglobin of 0, 0.3 and 0.6 g/L and tested them for cardiac troponin I (Beckman Coulter AccuTnI), cardiac troponin T (Roche Diagnostics), brain natriuretic peptide (BNP), and NT-prohormone-brain natriuretic peptide (NT-pro BNP) in modestly hemolyzed samples (i.e,. cell-free hemoglobin <0.6 g/L). Florkowski et al. evaluated the degree of analytical interference from hemolysis with currently available troponin assays [139]. Hemolysates of RBCs were prepared according to the original procedure reported by Meites [140]. Basically the investigators added various amounts of RBC hemolysates to four unhemolyzed pools of heparinized plasma with troponin I concentrations of 0.00, 0.03, 0.5, and 5 µg/L to produce a range of hemolyzed plasma samples. The hemoglobin concentration was measured as HI (calibrated in milligrams per deciliter of hemoglobin). Troponin I was assayed on both the Abbott Architect instrument – which displays a 99th percentile of the upper reference limit of the normal population of 0.28 µg/L and a 10% imprecision of 0.032 µg/L – and on the Ortho Clinical Diagnostics Vitros ECi instrument, which has a 99th percentile of the upper reference limit of the normal population of 0.034 µg/L and a 10% imprecision of 0.034 µg/L. Troponin T was instead assayed with both the Roche Troponin T STAT assay (99th percentile of the upper reference limit of the normal population of <0.01 µg/L and a 10% imprecision of 0.03 µg/L) and the highly sensitive Roche Troponin T assay (99th percentile of the upper reference limit of the normal population of 13 ng/L and 10% imprecision of 12 ng/L) on the ElecSys 2010 instrument. The within-batch coefficient of variation (CV) was <3% for all pools without added hemolysate. Nineteen percent of 966 consecutive samples collected in the ED were hemolyzed (HI >0.35 g/L); among these

12% were slightly hemolyzed (HI, 0.36 to 1.00 g/L), 4% were modestly hemolyzed (HI between 1.01 and 2.50 g/L), and 3% were grossly hemolyzed (HI >2.50 g/L). Both of the Roche assays displayed negative interference (up to −50%) with increasing degrees of hemolysis, whereas the Vitros ECi TnI assay showed a positive bias (up to 576%). Finally, the Abbott Architect TnI assay appeared to be more tolerant of interference from hemolysis (<10%).

In a subsequent investigation, Bais prepared hemolysate from RBCs that had been washed three times with 0.9% saline. After the final centrifugation, the cells were diluted with an equal volume of distilled water, thoroughly mixed, and lysed by freezing [141]. Troponin I was assayed with the Ortho Clinical Diagnostics TnI ES assay on a Vitros 5600 Integrated System, whereas troponin T was determined with the highly sensitive Roche assay on the Elecsys E170 immunoassay system. Thawed hemolysate was spiked to lithium heparin plasma samples that contained concentrations of cardiac troponin selected to be around the 99th percentile cutoffs for the respective assays − that is, 34 ng/L for troponin I (24, 36, and 49 ng/L) and 13 ng/L for troponin T (6, 12, and 23 ng/L). A 20% change was considered clinically significant, according the recent recommendation of the National Academy of Clinical Biochemistry (NACB), which indicates that this limit represents a significant (more than three standard deviations [SDs]) variation on the basis of a 5% to 7% analytical imprecision for discriminating the timing of injury as well as being suggestive of acute myocardial infarction that is either evolving (value increase) or resolving (value decrease) [142]. A HI of around 150 (i.e., free hemoglobin concentration of 1.9 g/L) caused a greater than 20% change in both assays. Therefore it can be concluded that contemporary high-sensitivity troponin assays are sufficiently affected at relatively low degrees of hemolysis to indicate that interference must be monitored for every specimen and results potentially suppressed to prevent misleading clinical decision making.

Overall, the typical decrease of troponin I concentration in hemolyzed samples has been attributed to enzymatic degradation by cardiac and extracardiac proteases, especially calpains, caspases, catepsin L, and gelatinase A, which might finally impair the immunoreactivity of the molecules. The heterogeneous results of the different studies on troponin I and T have been attributed to a variety of factors, including the epitopes recognized by the antibodies in the immunoassay, the sample matrix (i.e., serum or plasma), the type and concentration of the proteases, as well as the time elapsed before separation of serum or plasma from the cells.

A series of hemolysis interference studies were also carried out by Snyder et al. with the Vitros ECi (Ortho), and Elecsys (Roche) immunoassay platforms routinely used in the laboratory [130]. The authors observed that hemolysis caused a significant interference for 5 of the 16 assays evaluated using the Vitros ECi, with both PSA and troponin I assays showing a large increase in test results even in the presence of moderate hemolysis (for both tests the effect was dependent on the HI as well as on the analyte concentration, since the greatest positive bias was observed for samples with the lowest concentrations of both analytes). Assays for vitamin B12, testosterone, and cortisol exhibited a negative bias with increasing amounts of hemolysate and at all analyte concentrations. The effect was greatest on the vitamin B12 assay, but all three analytes showed an approximately 20% decrease in signal when gross hemolysis was present. The hemolysis interference was assessed on the Roche Elecsys for five immunoassays, including troponin T Stat, CK-MB Stat, beta human chorionic gonadotropin (βhCG), n-terminal pro–brain natriuretic peptitde (NT pro-BNP), and parathyroid hormone (PTH). Among these, only the troponin T assay exhibited clinically significant changes (i.e., the concentration of troponin T decreased with increasing hemolysate, becoming clinically significant with grossly hemolyzed samples). Folate is also present in erythrocytes at concentrations approximately 30 times greater than in serum; hence folate should not be assayed in samples with any degree of hemolysis [143]. Steen et al. reported clinically significant interference due to hemoglobin on ferritin and folate [144]. Cook et al. also reported a concentration-dependent negative bias in the measurement of insulin associated with increasing degrees of hemolysis on the Beckman Coulter Unicell DXI 800 [145].

More clinically interesting perhaps is the study by Kwon et al., who investigated the influence of hemolysis on the measurement of cardiac biomarkers (i.e., serum CK-MB, troponin I, and myoglobin) by adding an arbitrarily made interferent (RBC lysate) to decide on the acceptability of a specimen for these tests. Expectedly, CK-MB and troponin I were substantially affected by any degree of hemolysis, whereas myoglobin was mostly unaffected. It was therefore concluded that the results of cardiac markers should be carefully interpreted when the specimens are hemolyzed [146].

Verfaillie and Delanghe performed an elegant study assessing the correlation between increasing degrees of hemolysis and neuron-specific enolase (NSE) by analyzing a dilution series of pure hemolysate in water with a HI from 0 to 100 (or 0 to 33.2 g/L hemoglobin) [147]. The results revealed a nearly perfect linear correlation between the concentration of NSE and the HI (Hr^2 = 0.999). Moreover, a mean value of 0.30 µg/L NSE was found to be released by the RBCs per unit of HI. This consistent effect of hemolysis on the measurement of NSE prompted the authors to develop a compensating factor according to the degree of hemolysis present in the sample. Therefore adjustment of the NSE concentration was achieved by subtracting 0.30 µg/L per each HI unit from the measured NSE concentration. Hemolysis has also been reported to exert a negative bias in fluorescence polarization immunoassays [148] and some immunonephelometric methods, especially in assaying haptoglobin [149]. Finally, Moalem et al. have demonstrated that the values of intraoperative parathyroid hormone (IOPTH) in hemolyzed specimens might be decreased from 24.5% to 53.8% as compared with non-hemolyzed controls. Accordingly, unrecognized hemolysis in preexcision specimens could result in false-negative results and lead to unnecessary continued exploration, whereas unrecognized hemolysis in postexcision specimens might determine false-positive IOPTH results and lead to failed parathyroidectomy and the need for reoperation [150].

The pattern of protein electrophoresis might also be strongly modified in hemolyzed samples, since the hemoglobin-haptoglobin complexes move between the α2- and β-globulin fractions and free hemoglobin migrates as a diffuse reddish band in the β-globulin fraction. The result is a failure to separate α2- and β-globulin, which also leads to an underestimation of the albumin fraction.

8.4 Interference in ABO type and antibody screen testing

Transfusion errors are regarded as a serious clinical problem, since they dramatically jeopardize patient health and impose a huge economic burden on the healthcare system. The large majority of these errors are attributable to avoidable mistakes at various steps in the transfusion chain and – in analogy with the conventional laboratory diagnostics – largely prevail within the preanalytical phase of testing [151]. Rejection of hemolyzed specimens referred for ABO type and antibody screen can also occur in blood bank laboratories owing to concerns over spurious interference in assay performance and despite the current use of EDTA-anticoagulated specimens (i.e., magnesium is chelated by EDTA, which prevents fixation of complement and in vitro erythrocytolysis).

Tanabe et al. carried out a prospective cohort study of ED and labor-and-delivery patients to identify the factors affecting the risk of hemolysis in blood bank specimens [152]. The overall risk of hemolysis during blood collection was 7%, the risk of obtaining hemolyzed blood for testing was dramatically increased using Vialon™ IV angiocatheters as compared with steel needles (10% vs. 1.5%, respectively). The factors most significantly associated with hemolysis in a multivariate analysis were the use of Vialon IV catheters and sampling from an anatomical site other than the antecubital area.

Laga et al. investigated whether hemolysis produced via experimental mechanical stress (i.e., by applying the blade of the tissue homogenizer directly into the primary uncapped collection tube at speed setting 2 for 30, 60, and 75 seconds immediately after collection) has a threshold for introducing bias in ABO-Rh typing and antibody screening [153]. It was observed that although a significant proportion of blood samples collected from healthy human subjects and from patients with one or more alloantibodies became uninterpretable after mechanically induced hemolysis, in no case did a positive antibody screen become negative, nor was the reverse blood type changed. Those specimens that became uninterpretable after hemolysis had, however, been stored for a much longer period (i.e., 45 days vs. 12 days). Surprisingly, none of the samples from healthy human subjects with blood type O became uninterpretable at any given free supernatant hemoglobin concentration. It was therefore concluded that hemolyzed specimens referred for ABO-Rh typing and antibody screening are suitable for testing in the large majority of cases, so that recollection can be avoided. Only in the presence of gross hemolysis might a second sample be requested while the first one is processed for ABO-Rh front type, thereby minimizing the number of repeat venipunctures and sample collections.

9 Spurious hemolysis and veterinary medicine

9.1 Blood collection from small animals and pets

Hemolyzed specimens are also a rather frequent occurrence in veterinary medicine, perhaps understandably so owing to the challenge of collecting such blood, especially from small laboratory animals (e.g., mice) and pets (e.g., cats and dogs). Lewis et al. originally proposed an original technique to collect suitable amounts of blood from mice with as little hemolysis as possible [154]. In short, blood was obtained from a tail incision made immediately after the mice had been exposed to an ambient temperature of 45°C. Individual samples sufficient for most tests were readily obtained without significant hemolysis. The technique was also proven to be rapid and humane, requiring no special skill.

Suber and Kodell evaluated the effect of three phlebotomy techniques (i.e., periorbital sinus puncture, tail vein incision, and cardiac puncture) on hematological and clinical chemical parameters in Sprague-Dawley rats. The mean RBC and leukocyte count as well as the values of both hemoglobin and hematocrit were reduced with cardiac puncture as compared with the other two techniques. Moreover, the mean values of serum LDH, AST, ALT, GGT, and creatinine were greater in samples collected by cardiac puncture than in those collected by the other two techniques. Even more importantly, more than 60% of the serum samples obtained by using the cardiac puncture were hemolyzed, as compared with only about 25% of those obtained via tail vein incision. It was therefore concluded that the periorbital sinus venipuncture technique is the method of choice for collection of blood samples from rats because of the lower likelihood of obtaining hemolyzed specimens and the lower variance [155]. These results were, however, challenged by Aasland and colleagues, who concluded that saphenous venipuncture should be considered the method of choice in mice, since it carries a number of advantages when correctly performed [156]. Christensen et al. also assessed four methods of withdrawing blood samples from mice (i.e., amputation of the tail tip, lateral tail incision, puncture of the tail tip, and periorbital puncture) [157]. The presence of clots was recorded and hemolysis was monitored spectrophotometrically at 430 nm; another useful criterion was to verify whether it was possible to collect 30 to 50 μL of blood. All the methods were acceptable in terms of amount of plasma, and clots were observed in a limited number of samples with no significant differences between the methods. The periorbital puncture was not associated with any hemolyzed samples, the lateral tail incision resulted in only a few hemolyzed samples, whereas puncture or amputation of the tail tip induced hemolysis in a significant number of specimens. Therefore lateral tail incision was deemed to be the method of choice, since it is more likely to produce a clot-free and non-hemolyzed sample.

9.2 Interference of hemolysis in veterinary laboratory medicine

Although there is reliable information on the interferences caused by free hemoglobin in hematological or biochemical analyses and all manufacturers provide general advice on proper sample collection, less is known about the potential effect of this free hemo-globin on commonly determined analytes in animal serum samples. Martínez-Subiela and Cerón determined the effects of hemolysis on the most commonly used assays for C-reactive protein and serum amyloid A and the determination of ceruloplasmin values in dogs [158] by adding solutions of hemoglobin to serum aliquots. Hemolysis inter-fered significantly with test results for C-reactive protein and ceruloplasmin – but not with those for the serum amyloid A assay – the magnitude of the differences caused by the interfering substance did not appear to have an important impact on the clinical interpretation. Lucena et al. investigated the potential interference of hemolysis on commercially available enzyme-linked immunosorbent assays for cortisol and free thyroxine (FT4) in canine plasma samples [159]. Serum samples from 20 clinically nor-mal dogs were enriched in vitro with different amounts of fresh hemolysate and com-pared with the original sera. The presence of free hemoglobin significantly interfered with the accuracy of FT4 determination independent of hemoglobin concentration. In agreement with human samples, the presence of hemolysis in canine sera produced a change in electropherogram morphology, giving an interference peak located in the beta-2 region [160].

10 Detection of hemolyzed specimens

Error prevention, identification, monitoring, and management are the cornerstones of patient safety, but it seems challenging indeed to put all of these actions into practice and even more difficult to integrate all of them within a quality system. In recent years, a multitude of technological advances have made it possible to improve efficiency and confidence in the analytical phase of the total testing process. Therefore modern laboratory instruments streamline a variety of clinical chemistry and immunological tests; today's laboratory is also characterized by menu flexibility, high throughput, continuous workflow, on-board autodilution, fast turnaround times, and – last but not least – very accurate and precise measurements. Further important advances include the ability to process primary (closed) tubes, thus minimizing the risk of transmitting blood-borne pathogens as well the availability of on-board sample and reagent refrigeration to ensure specimen integrity, reagent stability, and reduced reagent waste. To limit inappropriate aspiration errors in samples with low or insufficient volume or in those containing microclots or bubbles, modern laboratory systems are now equipped with liquid-level sensors, so that electrical noise and accidental contact of the sensor with the tube or cup walls produces an alert. Both prevention of "carryover" and minimization of the dead volume for reagents and samples are additional advantages of these modern sensors.

The benefits of increasing automation originate mainly from replacement of manual, potentially dangerous, and error-prone steps with automated activities requiring minimal (or no) intervention from the operator [8]. It is useful to mention here, however, that the increasing infrastructural and organizational complexity has not facilitated the prevention and identification of preanalytical errors, because the integration and consolidation of preanalytical work stations with analytical platforms and even refrigerated storage units frequently conceal specimens from visual scrutiny. Detection of unsuitable specimens is actually impossible in analyzing whole blood specimens on POC devices. Recognition of unsuitable samples, especially those that are hemolyzed, is instead essential, since (a) they might indicate life-threatening disorders due to in vivo hemolysis that would require urgent notification of the clinicians, (b) laboratory testing on spuriously hemolyzed specimens is unreliable if not misleading, as described previously, and (c) discrepant results between POC devices and conventional laboratory instruments (e.g., for potassium) might be due to undetected in vitro hemolysis. Therefore the most debated issue regarding the overall quality of laboratory diagnostics remains the identification of unsuitable samples by the laboratory staff. Basically, hemolysis in serum or plasma is virtually undetectable until centrifugation of the specimen has been completed.

Even so, there are no definitive guidelines, other than recommendations or suggestions, defining unequivocal behaviors to assess the degree of hemolysis or establish widespread and universally agreed thresholds that should guide sample rejection [3,4]. Hemolyzed specimens have traditionally been identified on an arbitrary basis by visual inspection of the laboratory personnel. Nevertheless, this practice has been strongly discouraged for a variety of reasons. In a study by Glick et al., the frequency of turbid, hemolyzed, and icteric specimens was assessed in an acute-care general hospital by visual assessment using full-color photographs of sera adulterated with the three different interferents, which were designated as "0" (no interferent), trace, 1+, 2+, 3+, 4+, or 5+ [161]. The authors demonstrated that even trained observers were unable to accurately rank the degree of interference in serum even when they had a good benchmark, such as the photograph. In a further study, Hawkins assessed the agreement between visual grading of sample hemolysis with automated HI measurement on a clinical chemistry analyzer [162]. Photographs of mild (hemoglobin 1 g/L), moderate (2.5 g/L), and severe (5 g/L) hemolysis were used as a reference. The weighted kappa coefficient for serum samples was 0.42 with an 8.0% hemolysis rate determined by visual inspection and 3.4% by HI. For EDTA-plasma samples, the weighted kappa coefficient was 0.35 with a 4.5% hemolysis rate determined by visual inspection and 7.9% by HI, thereby confirming that visual assessment of sample hemolysis is unreliable and can also produce highly variable results depending on sample type. In a more recent investigation, Simundic et al. compared visual (i.e., comparison with photographs of samples containing various concentrations of hemoglobin) and automated detection (i.e., LIH reagent on Olympus AU2700 analyzer) of lipemia, icterus, and hemolysis in 1,727 routine biochemistry serum samples [163]. It was again confirmed that visual inspection was inferior to automated detection owing to the poor interrater agreement in estimating the degree of interference between laboratory personnel (mean kappa coefficient of 0.617). Finally, Jeffrey et al. compared the detection of hemolysis in adult and neonatal samples by inspection and measurement of HI [164]. It is also noteworthy that the presence of icterus dramatically decreased the ability to detect hemolysis on inspection, so that the use of HI rather than visual inspection is particularly advisable in the case of neonates whose serum tends to be icteric. Another potential source of interference is Patent Blue V, an inert dye increasingly being used during cancer surgery to identify the sentinel lymph node. Darby and Broomhead, in fact, observed a significant positive interference by 7.3 mg/L of Patent Blue in the lipemic index, as well as a significant negative and nearly linear dose-response bias in both the hemolysis and icteric indices measured on the Roche Modular, the former parameter showing a mean decrease of 71 ± 9.7 [165].

Although the quantification of free hemoglobin in serum or plasma might also be feasible using immuno-nephelometrical assays, the use of this approach on all samples referred to a clinical laboratory is impractical (increase in turnaround time, unavailability of the assay on all instrumentation) and prohibitively expensive. Along with other technical innovations described previously, several preanalytical work stations and laboratory instruments have thus been equipped with systems that are capable of automatically testing and eventually correcting for a broad series of analytical interferences (the "serum indices"), which also include hemolysis detection [166]. In most cases the instruments provide a qualitative or quantitative measure (i.e., HI), which must be compared with manufacturer-, instrument-, and analyte-specific alert values before a decision is made to perform or omit testing. Understandably this measure is not intended for diagnostic purposes but is used to obtain information on sample conditions. Therefore the operator can adjust levels at which the interference generates an alert, customizing as well the operating mode to reflect his or her personal operating requirements for reporting interference.

Although some disparities exist among the different manufacturers, the serum indices are usually obtained by monitoring serum or plasma absorbance at various wavelengths (traditionally between 340 and 670 nm) (▶Fig. 8.1). By solving a set of predefined equations, each index is computed and is directly proportional to the sample condition, as for icterus (bilirubin), hemolysis (free hemoglobin) and lipemia (turbidity) [167]. It is clear that the widespread implementation of the HI carries several benefits for daily laboratory practice. First, it is an objective, easy, rapid, and relatively inexpensive way to help laboratories standardize a streamlined process for the identification of hemolytic specimens, thus overcoming the inherent limitations of visual scrutiny, which is obviously subjective, arbitrary, and unreliable, especially in the "gray zone" between 0.2 and 0.3 g/L of free hemoglobin, where most specimens might appear clear (i.e., non-hemolyzed) at scrutiny. The automatic detection and subsequent rule-based algorithm for the actions to be undertaken can also remarkably increase the rate and effectiveness of detection for both serum interferences and in vivo hemolysis, so that routine reporting of the HI along with the results of single analytes should be considered. In the developing process toward implementation of sentinel events and quality indicators (i.e., to assess and monitor the quality systems of clinical laboratories) [168,169], objective measures of sample quality such as the HI offer the benefit of quality management, providing a suitable background for the development of notable and intolerable interference-induced bias. Finally, for instruments providing quantitative or semiquantitative results, serum indices can obviously be employed as a reliable measure of the degree of such interference and for the potential (mathematical) correction of test results, although we generally advise against this solution.

The use of the HI represents a crucial step for reducing the adverse outcome of laboratory errors as well as for harmonizing practices, two recent studies provide further remarkable insights on this promising tool. In the study of Söderberget et al., the HI was systematically used as a measure of the overall preanalytical quality of blood samples received in the laboratory from primary health care centers, nursing homes, and a hospital emergency department. Although the authors chose the lowest detectable HI level [170], which cannot be considered an universal criterion for sample rejection, they should be praised for their efforts, since they were able to demonstrate the existence of significant variations in the prevalence of mildly hemolyzed samples among the various referral units, also identifying suboptimal preanalytical conditions for collection and handling of the specimens across the various healthcare facilities. In particular, the authors reported that samples from the primary healthcare center with the highest prevalence of hemolysis were 6.1 times more often hemolyzed as compared with those from the center with the lowest prevalence. Of the samples collected in primary health care, 10.4% were hemolyzed, as compared with 31.1% of those collected in the ED. A remarkable difference in hemolyzed samples was also observed between the ED section staffed by emergency medicine physicians and the section staffed by primary healthcare physicians (34.8% vs. 11.3%, p <0.001). According to these data, the HI was proven useful for both checking the integrity (quality) of the specimens and also for estimating (and monitoring) preanalytical quality (i.e., skills and experience of the personnel with sample collection responsibilities). In a second study, a multicenter investigation on the HI involving different clinical chemistry instruments (Roche Modular System P, Siemens Dimension RXL. Siemens ADVIA 2400 and 1800, Olympus AU680, and Beckman Coulter DXC 800), we showed that most of the instruments tested provided comparable HI results. The quantitative HI results obtained on the four Modular System P units included in the study were also highly reproducible, as attested by a nonsignificant variation ($p = 0.911$ by the Kruskal Wallis test) and an excellent correlation by Passing and Bablok regression analysis and Pearson correlation coefficient. Moreover, in analyzing the results of instruments providing quantitative HI results, the mean intra-assay CVs calculated on triplicates were excellent, ranging from 0.1% to 2.7%. Nevertheless, it was also emphasized that results reporting (i.e., quantitative or semiquantitative data, measure unit) as well as manufacturer-specific thresholds were rather heterogeneous, so that samples characterized by borderline hemolysis (approximately 0.5 g/L) might have been considered still suitable on some analytical platforms but not others [171].

As regards specimen type, Unger et al. [174] collected blood from 49 patients into sample tubes containing EDTA (1.6 g/L of whole blood) or Li-heparinate (16 kIE/L of whole blood). Serum indices were assessed on the P-module of the Modular system (Roche), while free hemoglobin was assessed with the two-wavelength method of Golf et al. [172], which is based on differences in absorbance at 540 and 600 nm. A hemoglobin solution was used as a calibrator at 1.25 g/L. The method detection limit was 0.05 g/L and at concentrations of 0.09, 0.3, 1.25, and 9.5 g/L, intra-assay imprecision values (CVs) for HI were 15%, 3.6%, 0.6%, and 0.3%, respectively. Interassay CVs, determined by replicate measurements during 10 days, were between 0.6% and 12.4%. In univariate analysis of variance (ANOVA), no significant differences were observed between specimens collected in EDTA or heparin regarding both free hemoglobin and the HI. Moreover, the correlation coefficients between free hemoglobin and HI were 0.939 and 0.967 in EDTA and Li-heparinate plasma, respectively. Difference plots for both plasma collections showed no correlation of difference with the mean of free hemoglobin and HI. Hence, although the previous notion provided by Thomas and Thomas [173] that free hemoglobin may be 20-fold higher in EDTA than in heparin plasma, could not be confirmed, a bias from EDTA became apparent when blood from healthy volunteers was collected with increasing EDTA concentrations. More specifically, both free hemoglobin and the HI displayed more than 50% higher values at EDTA concentrations three times higher than those of EDTA samples collected under standard conditions. It is noteworthy that such an effect was not observed in Li-heparinate plasma, which showed no effect on both free hemoglobin and HI quantification in the Li-heparin range of 16 to 48 kIE/L. It was hence concluded that EDTA might influence both free hemoglobin and the HI in a concentration-dependent manner, but this bias is observed only when tubes are not correctly filled to their nominal volume [174].

Regardless of some technical differences, further efforts toward standardization or harmonization of this useful measure should be planned, such as the document "Use of Serum Indices (Hemolysis, Icterus, Turbidity) in the Clinical Chemistry Laboratory," which should be regarded as a valuable enterprise being developed by the CLSI [175]. It is also worth mentioning that the spectrophotometric detection of the HI cannot be regarded as a universal remedy and some practical considerations about its routine implementation should be considered. Although the quantification of the HI is easy and relatively inexpensive, the multiple benefits previously mentioned should still be weighted against the potential delay of workflow of samples and the potential prolongation of the TAT, especially in high-volume and STAT laboratories, so that additional investigations assessing the impact of serum indices on laboratory activity are also advisable.

We firmly believe that the implementation of HI is a paradigmatic example of how the partnership between laboratory professionals and the in vitro diagnostic industry can increase the reliability of testing and also enhance patient safety [176]. Strong support to the widespread introduction of this valuable instrumental tool for assessing sample quality as well as preanalytical practice comes from the recent study by Koseoglu et al., who observed that hemolysis interference affects LDH and AST at almost undetectable levels of hemolysis by visual inspection (i.e., plasma hemoglobin below 0.5 g/L) [108]. It is also noteworthy that some clinical laboratories have actually incorporated the HI into their reporting algorithm and occasionally also within their laboratory reports, thereby allowing complete automation of both detection and processing. As an example, Vermeer et al. developed an algorithm for the detection and processing of clinically or analytically relevant amounts of hemolysis, icterus, and lipemia in several laboratory assays [177]. The decision rules and pertinent activities were first introduced, and the algorithm was then implemented in the laboratory information system (LIS). Before the laboratory professional releases the test results, each assay is validated according to the minimal concentration of free hemoglobin that would induce a bias (either positive or negative), so that the result can be managed according to predetermined procedures. Thus if interference is present below a specific critical concentration, the result is automatically accompanied by a warning comment to alert the clinician. When the index exceeds the critical concentration, the result is instead flagged, so that the technician is informed that the result cannot be reported to the clinician. It is noteworthy that the introduction of systematic assessment of serum indices and decision rules has increased the detection rate of relevant hemolysis 6-fold, whereas the introduction of the algorithm in the LIS further increased the detection rate for relevant serum interference 69-fold.

Unfortunately, however, there is clear evidence that the valuable tool reflected by the HI is still largely underused. In the EPSC–IFCC WG-LEPS survey, labs were asked "how do you check to evaluate sample hemolysis in your lab?" The reply from 56% of the labs was "visual inspection," 43% replied by "measuring the HI," and 1% answered "we do not check." In addition, the labs were asked "Do you systematically monitor the number and the origin (wards, facilities, etc.) of hemolyzed specimens that you receive in your lab?" In reply, 42% of the participants said no and 58% said yes (▶Tab. 4.2).

A specific challenge is represented by whole blood specimens, as, for example, referred for blood gas analysis and hematological and POC testing. These samples are typically processed without first being centrifuged, so that the presence of hemolysis is not routinely assessed or easily assessable. Nevertheless, Lippi et al. suggested that systematic and rapid centrifugation (e.g., at 3500 × g for 5 minutes) after the requested analyses were completed, followed by scrutiny at least by visual inspection (since HI analysis might be unsuitable because of the negative impact on the turnaround time), might be a reliable strategy to establish whether or not the samples contained free hemoglobin and were suitable for testing [18].

11 Management of hemolyzed specimens

Regardless of the underlying causes, the potential rejection upon receipt of hemolyzed specimens can lead to inconvenience and delay in clinical decision making, with substantial implications for the care of patients. Moreover, an additional cost is incurred when specimens must be recollected. Since any repeated collection of a blood tube costs around $0.10 (US) and requires 0.25 minutes plus repeat or additional testing, the overall additional cost in terms of laboratory operations can become considerable.

The issue of hemolyzed specimens has traditionally plagued clinical laboratories, urgent measures for developing operative guidelines and recommendations are required, given the evidence that there is no standard way of dealing with this problem worldwide as yet . As for any other type of medical error, the implementation of a total quality management system is hence the most effective strategy, encompassing a multifaceted strategy for process and risk analysis based on error prevention, detection, and management. The recommendations should, however, always be used in conjunction with clinical signs and knowledge of the underlying disease processes as guides for interpreting laboratory data and determining the accuracy of reported laboratory results.

11.1 Local evaluation of the bias

The foremost issue is to define reliable limits for establishing whether a certain degree of hemolysis in the specimen would produce a bias in the local test system (▶Tab. 11.1). There are comprehensive data on this issue, both from the current literature and from the manufacturers' reagent package inserts, the best solution is still to redefine such limits locally, since many inserts contain limited information on interferences and often only on what concentration of material would interfere with the assay and with no information on what concentrations of analyte were tested. Moreover, the bias typically depends on a wide range of variables, including (a) the sample matrix (e.g., serum versus plasma versus whole blood); (b) the concentration of free hemoglobin in serum or plasma (i.e., low vs. degree of hemolysis); (c) the test being performed (e.g., potassium vs. immunochemistry testing); (d) the reagents being used (e.g., for creatinine, Jaffe vs. enzymatic determination); and (e) the analytical instrument. Moreover, manufacturers' technical disclaimers regarding the potential effect of an "interferent" are often ambiguous or reported as "hemolysis should be avoided," "hemolysis might interfere," and so on. It is instead useful to gather specific data on the qualitative and quantitative bias of a given interferent on the analytical system in use. Therefore, since the package inserts are often insufficient and the findings of individual studies on this topic cannot be generalized or translated from one assay condition to another, the potential bias introduced by any "interferent" should be considered highly variable and investigated on a local basis. This can be accomplished most reliably by spiking plasma or serum samples with serial dilutions of free hemoglobin, performing the analysis, and then identifying the cutoffs (i.e., free hemoglobin values) where the bias is to

Tab. 11.1: Assessment of the interference of hemolysis

- Prefer objective and validated methods (i.e., the hemolysis index) to visual inspection for identifying hemolyzed specimens.
- Consider reporting the hemolysis index (HI) on laboratory reports.
- Describe clearly the cutoff of hemolysis (free hemoglobin) after which the interference might be clinically significant for each laboratory test.
- Perform local experiments to document the interference on the methodology in use, as follows:
    1. Generate a hemolysate by means of:
        a. Overnight freezing and thawing
        b. Washing four times with cold isotonic saline solution
        c. Washing four times with deionized water and saponin
        d. Passing the specimen through a blood collection needle
        e. Stirring with a metallic bar for 1 to 5 minutes
        f. Applying the blade of a tissue homogenizer directly into the uncapped primary collection tube
        g. Adding purified human hemoglobin (human isoionic oxyhemoglobin)
        h. Sonication
    2. Obtain human fresh reference serum or plasma specimens to which the hemolysate should be added and which should be characterized by the concentrations of analytes covering both normal and pathological values.
    3. Produce a scalar dilution of the hemolysate, which should cover the range of concentration of free serum or plasma hemoglobin conventionally encountered in the laboratory, where this "hemolysis interferograph" is supposed to be used.
    4. Use objective criteria (e.g., the total allowable error or the analytical quality specifications for desirable bias or the reference change value) to identify the bias of results in hemolyzed specimens.

---

be considered clinically significant, either when the bias exceeds the desirable specifications for imprecision, inaccuracy, total allowable error calculated from data on within-subject and between-subject biological variation, or when it significantly surpasses the limit defined by the critical difference for any analyte tested. In short, hemoglobin should be added to serum with the least possible dilution, so that it simulates the hemolysis created by erythrocyte destruction under the conditions of blood collection or as it originates from blood clotting, processing, or treatment.

11.1.1 Preparation of the hemolysate

There are several approaches to preparing and then testing the bias of "interferents" (including hemolysate) on laboratory testing. The simplest (i.e., freezing and thawing) is also the oldest, having been known and practiced since 1953 [178]. Nevertheless, although RBCs can be efficiently lysed by freezing whole blood or RBC concentrates at −10°C to −40°C for at least one minute, the complete lysis of all blood cells (i.e., including leukocytes and platelets) requires a longer period of freezing at temperature between −20°C and −40°C, preferably over 4 to 6 hours. In a recent investigation, Lippi et al. collected whole blood into siliconized vacuum tubes without gel separators containing 18 U/L lithium heparin; they prepared six aliquots, which were then frozen at different temperatures for varying periods [179]. The first aliquot was immediately analyzed, whereas the following ones were stored at −20°C or −80°C. The tubes stored at −20° C were thawed at 1, 2, 4, and 12 hours after freezing, whereas those stored at −80°C were thawed after 2 hours of freezing. Significant lysis of leukocytes and platelets occurred after 1 hour of storage at −20°C, the RBC count fell only modestly; consequently the HI of plasma rose only modestly from 0 to 3 (i.e., cell-free hemoglobin between 1.0 and 1.5 g/L). A greater degree of hemolysis occurred after 2 and 4 hours of storage at −20°C, since the RBC count decreased and the HI of plasma increased. The leukocyte count was instead only marginally reduced. Twelve hours of freezing at −20°C caused a sharp decay in both the RBC and white blood cell (WBC) counts. After 2 hours of storage at −80°C, most blood cells were also efficiently lysed.

Meites has suggested washing the packed erythrocytes obtained from heparinized whole blood three times with isotonic saline [140]. The cells are then diluted with an equal volume of water and frozen overnight. After being thawed at room temperature, the cells are centrifuged and the stromae removed. The hemoglobin concentration in the supernatant fluid is then determined with a reliable assay (e.g., cyanmethemoglobin). This "stock" hemoglobin solution contains nearly 90 to 110 g/L. Then, in the preparation of an "experimentally" hemolyzed specimen, aliquots of the stock hemoglobin prepared as in the previous paragraph are added to "normal" serum or plasma free of suspected interferences in variable volumes to obtain the desired amount of final free hemoglobin concentration in the "normal" serum or plasma specimen [140]. Glick et al. have suggested preparing a fresh hemolysate (hemoglobin content at least 200 g/L) by washing erythrocytes (from the same volunteers who were the source of serum or plasma) four times with cold isotonic saline solution. After resuspension of cells and repacking by centrifugation, the supernatant solution and leukocytes are discarded. After a final centrifugation to remove the saline solution, the erythrocytes are lysed with distilled water. The hemolysate is then filtered through untreated glass wool and centrifuged to remove cellular debris. The concentration in the final clear hemolysate is quantified by a cyanmethemoglobin assay. The hemolysate is finally added to serum or plasma samples, appropriately diluted with water, to produce a series of scalar concentrations [180]. At variance with this procedure of Glick et al., Frank et al. prepared hemolysates from EDTA-anticoagulated whole blood samples by washing the RBCs three times with isotonic saline and then adding an equal volume of deionized water and sodium saponin [98].

One disadvantage of some of the previously mentioned methods is that the hemolysate is largely free of hemoglobin. As previously noted, for some test systems (e.g., hemostasis assays), cell lysates and releasates other than hemoglobin may cause interference in testing. Accordingly, an alternative technique for reproducing in vitro hemolysis during the collection of blood more closely has been proposed by Dimeski. Briefly, whole blood anticoagulated samples are aliquoted and then passed through a blood collection needle to mimic a typical collection process. The number of times each sample is passed through a needle increases with each subsequent sample to produce a broad range of degrees of hemolysis [181]. The hemolysate is then added to serum or plasma in scalar dilutions to obtain a scale of interference, as described previously. Yücel and Dalva have also obtained hemolysate by mechanical trauma (i.e., stirring the whole blood with a metallic bar for 1 to 5 minutes). This method of lysis reliably mirrors the mechanical disruption of RBCs that might occur during faulty blood collection and sample preparation [102], enabling the production of three groups of specimens classified as being non-hemolyzed (mean free hemoglobin concentration of 0.012 g/L), moderately hemolyzed (free hemoglobin concentration of 0.11 g/L), and severely hemolyzed (free hemoglobin concentration of 0.28 g/L). Laga et al. hemolyzed the blood within primary uncapped collection tubes by applying the blade of a tissue homogenizer at speed setting 2 for 30, 60, and 75 seconds immediately after collection [153]; even more interestingly, they observed no difference between group O and non–group O blood as regards vulnerability to mechanical hemolysis. Finally, Milowska et al. verified whether ultrasound exposure might alter plasma membranes and the integrity of RBCs [182]. Ultrasounds were produced by an apparatus for supersonic therapy and applied to RBCs contained in small foil bags. Hemolysis was then assessed by measuring the hemoglobin content in the supernatant at 540 nm. Ultrasound at intensities of 1.90 and 2.44 W/cm^2 for 5 minutes induced a significant trend toward hemolysis as compared with the control samples (about 3.5% hemolysis) in those subjected to the highest intensity (about 5.1% hemolysis; $p < 0.001$). The control cells were less fragile and resistant to hypotonicity (i.e., dilution with media containing saline at concentrations from 0.6% to 0.15%) as compared with sonicated RBCs.

The strategy used by Blank et al., which includes spiking serum pools with increasing concentrations of purified human hemoglobin (human isoionic oxyhemoglobin), seems less suitable, since assessment of the bias would not take into account the additional interference generated by other intracellular components and cell debris produced from the breakdown of RBCs as well as leukocytes and platelets [92]. Irrespective of the technique used to produce the hemolysate, the choice of its lowest, highest, and intermediate concentrations should cover the range of concentrations encountered in the laboratory in which this "hemolysis interferograph" is supposed to be used.

Therefore the preferred procedure, at least the one that we can recommend, is based on mechanical trauma of whole blood by aspiration with an insulin syringe equipped with a very thin needle (e.g., ≤30 gauge). A first aliquot is readily separated from the whole blood and processed without further manipulation, since it is considered the "baseline," with no interferent. The further aliquots are then obtained by increasing aspirations of whole blood with the same syringe, as depicted in ▶Fig.11.1. This method reliably reproduces a traumatic blood collection with production of a poor-quality specimen and is also expected to damage platelets and leukocytes other than erythrocytes.

11.1.2 Measurement of hemoglobin in biological samples

Several techniques have been proposed to assess the concentration of free hemoglobin in biological specimens. The method of choice is that currently recommended by the CLSI (former National Committee for Clinical Laboratory Standards [NCCLS]) [183] using the hemiglobincyanide (HiCN) assay (conventionally and also known as "Drabkin's"). This method is deemed suitable for whole-blood calibration procedures for automated hematology analyzers in the evaluation of instruments and alternative methods for the determination of hemoglobin concentration and should be applied when patient RBC measurements are used for the calibration and control of hematology

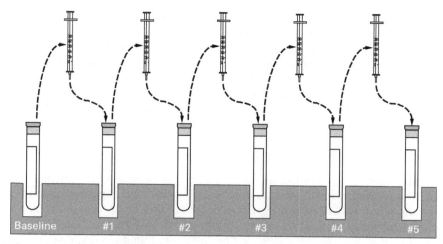

Fig. 11.1: Recommended procedure for generating hemolysis and blood cell lysis.

analyzers. This recommendation dates back to 1958, when the Panel on the Establishment of a Hemoglobin Standard, Division of Medical Sciences – National Research Council reviewed several photometric methods used for determining hemoglobin levels and concluded that, for a variety of reasons, the best method was that in which hemoglobin was measured after conversion to cyanmethemoglobin. Originally a colorimetric cyanmethemoglobin method was proposed by Stadie in 1920 [184], where total hemoglobin at alkaline pH is rapidly converted to the cyanoderivative, whose absorbance was then measured at 540 nm. This method was then simplified by Drabkin and Austin in 1935 [185] by combining the separate reactants alkaline ferricyanide and cyanide into a single reagent. Basically Drabkin's assay is based on the oxidation of hemoglobin and its derivatives (except sulfhemoglobin, a pigment that normally occurs in only minute concentrations in blood) to methemoglobin in the presence of alkaline potassium ferricyanide. Methemoglobin then reacts with potassium cyanide to form cyanmethemoglobin, which has maximum absorption at 540 nm. The color intensity measured at 540 nm is proportional to the total hemoglobin concentration. The color is suitable for measurement in filter as well as in narrow-band spectrophotometers because its absorption band at a wavelength of 540 nm is broad and relatively flat, thereby permitting measurements using both wide- and narrow-bandwidth instruments (530 to 550 nm). The addition of specific surfactant also minimizes the turbidity occasionally caused by the presence of RBC stromae.

In 1963, the Standardizing Committee of the European Society of Haematology became the International Committee for Standardization in Haematology (ICSH), and the ICSH Expert Panel on Haemoglobinometry was formed to draw up recommendations about hemoglobin testing. These were accepted at the International Congress of Haematology in 1966 and published in 1967. Meanwhile, the National Institute of Public Health in the Netherlands prepared and made available, on behalf of the ICSH, an international hemiglobincyanide (HiCN) reference solution, one lot of which was accepted by the World Health Organization (WHO) as the International HiCN Reference Preparation (WHO Techn. Rep. Series 384: 85, 1968). The WHO subsequently accepted further batches of HiCN reference solution as the second, third, and so on

International HiCN Reference Preparation (now the reference standard). The international HiCN reference solutions are controlled by laboratories in Italy, Japan, the Netherlands, the United Kingdom, and the United States. Although some other methods for the determination of hemoglobin have been described over the past decades (e.g., those based on estimation of oxygen, carbon monoxide capacity, or iron content), they proved to be less reliable because of the heterogeneous nature of hemoglobin, so that the Drabkin's reagent still remains the benchmark against which all other methods should be tested.

11.1.3 The "reference" serum or plasma specimens

The choice of the reference serum or plasma specimens to which the hemolysate should be added to test the bias is also critical. First, these might be human fresh samples and not quality-control materials in order to rule out any potential "matrix effect" when used with different instruments and test reagents. This recommendation is supported by the clear evidence provided by Miller et al. that the occurrence of noncommutable results for quality-control materials is frequent enough that the results cannot be used to verify consistency of results for patient samples [186]. Then, understandably, the range of concentration of any given analyte for which the interference is to be tested should include both normal and pathological values in order to verify the bias in both "health and disease." It is thus advisable to obtain at least three reference sera or plasmas with analyte concentrations below and above the limits of the reference range of the specific (local) assay as well a serum or plasma specimen with a "normal" value (e.g., for testing the interference of hemolysate on serum potassium, which is one of the most affected tests in hemolyzed specimens). It might be advisable to use three serum or plasma samples with potassium concentrations at 2.5 to 3.0 mmol/L ("control low"), 3.5 to 5.0 mmol/L ("normal control"), and 5.0 to 6.0 mmol/L ("control high").

11.1.4 Definition of the allowable bias

The next important steps include an evaluation into whether the presence of free hemoglobin in serum or plasma generates a bias in a certain assay as well as the relative amount of free hemoglobin in serum or plasma at which that test is "significantly" biased. As regards the statistical evaluation of data, linear correlation analysis might be helpful but does not provide a comprehensive representation of the bias, since a high correlation does not automatically imply that there is good agreement between two sets of data. Therefore a Bland-Altman plot (also known as a "difference plot") is the method of choice for data plotting, which is also used for analyzing the agreement between two different assays. It is basically identical to a Tukey mean-difference plot but was popularized in medical statistics by Bland and Altman [187–190]. Once the mean or median variation (i.e., the bias due to the presence of free hemoglobin in the specimen) has been calculated for all the analytes tested (in either absolute or relative terms), these must be compared with the "allowable" bias (i.e., the clinically significant variation).

Either the total allowable error (TE), or the analytical quality specifications for desirable bias, which are both derived from intra- and interindividual variations [191], as well as the specific reference change value is suited for this purpose. Total error represents the overall or total error that may occur in a test result owing to

The vast majority of laboratory errors are due to disorganization or lack of standardization, so that the implementation of guidelines or consensus recommendations for the detection and management of unsuitable specimens would be advantageous in that they help to ensure uniform operator behavior both within the same laboratory and among different laboratories. Improved communication among caregivers and interdepartmental cooperation is also essential for disseminating best practices and improving the quality of specimens. However, this process requires a thorough audit within the clinics and in the peripheral phlebotomy facilities. In fact, there is much better compliance with specimen collection policies and procedures when phlebotomists and specimen collectors clearly understand why things must be done in a particular way.

Along with the appropriate training of phlebotomists, a deep knowledge of the types of analytical and biological interferences that may affect hemolyzed specimens is also essential to minimize the chance of error at any step of the preanalytical phase. Besides the implementation of standardized collection procedures, additional aspects might have a role in decreasing the chance of obtaining unsuitable (e.g., hemolyzed) specimens, such as the use of conventional straight needles larger than 23 gauge along with evacuated tube systems. The use of evacuated tube systems is widely recommended, the nature of the patient care process in the ED does mean that there will be a justifiable need, in some circumstances, to collect blood using a syringe-based method. The use of large (e.g., 10 to 20 mL) sizes for syringe collections may be convenient, but it can generate excessive negative pressure leading to high flow rates (and thus high shear forces) that produce cell rupture. Smaller-size syringes (e.g., 3 to 5 mL) are hence recommended, along with a slow rate of draw on the plunger.

Practices and procedures should be carefully reviewed in collecting blood samples from subjects with small and/or fragile veins (i.e., neonates as well as pediatric, geriatric, and cancer patients). Collection from a hematoma site and prolonged tourniquet time should be avoided, as should equipment and connections that can lead to a turbulent blood path (butterfly devices or cannulas). Venous stasis should also be minimized, and the use of transillumination devices based on cold near infrared light-emitting diodes has been proposed to simplify both the location of the the vein and collection of blood in newborns and children as well as for mapping veins to be cannulated prior to ambulatory phlebotomy because it allows accurate visualization of the vein course, thereby eliminating venous stasis and improving quality control in the phlebotomy procedure [193,194].

Vigorous mixing of the specimens after collection should be avoided and appropriate conditions of temperature and humidity should be maintained throughout the specimen processing and handling phases. Defined and standardized practices for sample transportation and storage should be applied and the blood specimens be centrifuged within a reasonable time of collection in appropriate conditions (force, spin time, and temperature). Finally, the supernatant (serum or plasma) must be separated from the blood cells in a timely manner unless the primary tube is provided with a (gel) separator.

11.4 Preventing hemolysis from intravenous catheters

It is generally established that blood samples for laboratory tests should be obtained via venipuncture – preferably using vacuum tubes with straight needles because such needles provide a smooth, solid inner lumen surface that is unaffected by drawing pressure, which can produce spurious hemolysis – samples are occasionally collected using less suitable devices such as syringes, butterfly devices, small-gauge needles, and even IV catheters, especially in ICU and ED. Whereas sample collection using syringes is sometimes unavoidable, medium-size syringes (3 to 5 mL) are preferred to those that are much larger or smaller. It has also been endorsed that experienced ED nurses can reduce the number of hemolyzed specimens by collecting via venipuncture instead of through IV catheters, so that this practice should be considered as the standard of care in the ED setting. Unfortunately little scientific research exists on optimal methods for obtaining blood samples from catheters, and clinicians use a variety of unproven techniques. These are listed in ▶Tab. 11.3 [195].

11.5 Reliable identification of hemolyzed specimens

Another crucial step in the management of hemolyzed specimens involves their reliable identification and, in particular, to the definition of the degree of hemolysis that is most likely to affect the results of laboratory testing. As mentioned previously, modern laboratory technology might be equipped with software capable of automatically testing and eventually correcting for a broad series of analytical interferences, including the hemolysis index (HI). The use of this technology is advisable for several reasons, namely: overcoming the inherent limits of visual inspection, improving the recognition of mildly hemolytic specimens (i.e., those with free hemoglobin below 0.5/0.6 g/L) that are virtually undetectable by visual inspection but still unsuitable for some measurements, and standardizing and harmonizing practices among operators in the same laboratory and among different laboratories. It might also be worthwhile to report the HI on the laboratory report to either testify to the suboptimal quality of the specimen

Tab. 11.3: Recommendations for practice: collecting blood from intravenous catheters

1. Discard method
   - Flush catheter prior to obtaining specimen and use a discard specimen especially when collecting samples to be used for measuring drug levels.
   - Consider removing at least three times the catheter volume to clear the catheter of infusion liquid.
   - If removing discard with a syringe, use a new syringe (other than the discard or reinfusion syringe) to obtain the specimen.
   - Always label one tube or one syringe as "discard" prior to drawing a discard sample so as to avoid the risk of confusing the discard with an actual blood sample for analysis.
2. Reinfusion method
   - Methods of drawing blood requiring reinfusion of discard are discouraged because they may introduce clots into the system, although whether the clots present in the catheter and their reinfusion represent a significant risk to patient outcome is unclear.
3. Push-pull or "mixing" method
   - This procedure may be accurate for other than coagulation and drug levels and reduces blood loss, exposure risk, is easy to perform, and may be accurate for chemistry tests.

(when hemolysis is attributable to preanalytical problems) or to provide clinicians with a quantitative assessment of the free hemoglobin in blood (when hemolysis is due to in vivo causes). Since the serum indices are currently performed by most laboratory instruments as spectrophotometric assays, it seems reasonable that these tests should fulfill the same analytical requirements and be monitored in the same way as any other laboratory assay. Basically, quality control in laboratory diagnostics aims to detect, attenuate, and even correct deficiencies in a given analytical process, thus improving the quality of the test results. Quality control is hence a measure of both precision and accuracy and should display the same specific features, such as being simple to use, being derived from the same matrix as patient specimens (including viscosity, turbidity, composition, and color), displaying minimal vial-to-vial variability (variability could otherwise be misinterpreted as systematic error in the method or instrument), and being stable for long periods of time in large enough quantities. In short, internal quality control (IQC) and external quality assessment (EQA) (sometimes also referred to as proficiency testing) are two distinct yet complementary components of a laboratory quality assurance program. IQC is used to establish whether a series of techniques and procedures are performing consistently over a period of time. It is therefore deployed to ensure day-to-day laboratory consistency. EQA is used to identify the degree of agreement between one laboratory's results and those obtained by others. In large EQA schemes, retrospective analysis of results obtained by participating laboratories permits the identification not only of poor individual laboratory performance but also of reagents and methods that produce unreliable or misleading results. Taking these concepts for granted, in the foreseeable future both IQC and EQA should be extended to embrace the extra-analytical phase of the total testing process and therefore the serum indices, which also include the HI. The feasibility of such programs has already been proved in a multicenter study, where frozen human serum samples spiked with different amount of hemoglobin were shipped and analyzed across Europe [171]. The use of frozen human (serum or plasma) specimens for this purpose is also in agreement with recent evidence that both quality control and proficiency testing materials should be derived from fresh samples, since lyophilized materials have been reported to be noncommutable with clinical patient samples in comparing results between different measurement procedures [186].

Since hemolysis has a varying influence on different analytes, methods, and instruments, a detailed knowledge of each analytical technique and instrument is essential, so that each laboratory can clearly specify, within the local operative procedures (i.e., quality manuals), (a) the type of analyses that might be influenced and (b) the relative degree of hemolysis that is believed to alter test results. This is supposed to have already been provided by the manufacturer of the instrument/reagent, a further local assessment might be worthwhile.

11.6 Management of test results on hemolyzed specimens

After identification of a hemolyzed specimen and quantification of the degree of hemolysis, there are at least three potential approaches to the management of the test results: (a) correction of the data for the degree of hemolysis, (b) reporting of test results with an interpretive comment, and (c) warning clinicians of problems and the possible recollection of specimens. An alternative hypothesis – that is, diluting the sample to lower the burden of the interference – cannot be considered in this setting, since the bias in the measurement of most analytes (i.e., potassium, LDH, AST) is caused by intracellular release rather than by analytical interference. Thus this approach could be beneficial for only a limited number of parameters, such as total bilirubin [166].

11.6.1 Data correction for the degree of hemolysis

The first consideration in dealing with hemolysis interference is that those analytical techniques less influenced by hemolysis should be preferred. Then, suppression of the analytical interference may be challenging or impossible, it is theoretically feasible to modify the spectral interferences by multiwavelength analysis, blank measurement, or preparation before testing. A certain degree of analytical interference might still be eliminated, this approach does not apply for reporting results of those analytes where the final concentration is more influenced by intracellular leakage or dilutional effects. Some corrective factors for factitious hyperkalemia in the clinically relevant range of serum potassium concentrations have also been identified, so that adjustment of test results (especially potassium) for the degree of hemolysis has also been suggested as a reasonable step in order to provide clinically useful information on unsuitable specimens (e.g., with accompanying comments on the laboratory report such as "If a diagnosis of intravascular hemolysis can be excluded, in vitro hemolysis might have caused a false increase/decrease of . . ."). This approach is based on specific equations, where the hemoglobin concentration is multiplied by the slope (ranging from 0.0028 to 0.00319) obtained from a linear regression analysis between the bias observed for potassium at the relative free hemoglobin concentration in serum or plasma [196,197]. Vermeer et al. have also proposed some formulas based on the HI for correcting clinical chemistry test results, as follows: [corrected potassium = measured − (HI × 0.14)]; [Corrected LDH = measured LDH − (HI × 75)] [166]. Additional approaches have been suggested, where the potassium released by RBCs after hemolysis is derived from a formula including the MCHC along with free hemoglobin, as follows: potassium released by RBCs = (measured potassium + MCHC) × free hemoglobin [198], or where corrected potassium is calculated from the HI [164,166,199] (▶Tab. 11.4). Theoretically, when the lower bound of the predicted delta potassium still provides a roughly acceptable value (i.e., within the reference range), a second blood draw might be unnecessary. Such an approach might, however, be inaccurate and potentially misleading for

Tab. 11.4: Overview of some formulas used to adjust the concentration of potassium in hemolyzed specimens

---

- Adjusted potassium = free hemoglobin × 0.00319
- Adjusted potassium = free hemoglobin × 0.0028
- Adjusted potassium = measured potassium − (hemolysis index × 0.004)
- Adjusted potassium = measured potassium − (hemolysis index x 0.01)
- Adjusted potassium = measured potassium − (hemolysis index × 0.25)
- Adjusted potassium = measured potassium − (hemolysis index increment × 0.14)
- Potassium released by RBC = [measured potassium + MCHC] × free hemoglobin

---

Tab. 11.5: Main drawbacks of using corrective formulas to estimate potassium concentration in hemolyzed specimens

---

- Heterogeneity of the different formulas
- Heterogeneous release (i.e., broad bias) of potassium from hemolyzed red blood cells
- Misleading picture of the biochemical profile in patients with in vivo hemolysis
- Dependence upon the instrumentation and reagents in use

---

a variety of reasons, as listed in ▶Tab. 11.5. The first drawback is the huge number and the heterogeneity of the different formulas, which must be system-specific and highlight not only a substantial lack of agreement among the scientists but also the potential to obtain a wide spectrum of adjusted potassium results in spite of an identical concentration of free hemoglobin in the specimens. It is widely known that standardization is a key step for obtaining reliable and comparable results of laboratory testing and, in this specific circumstance, we seem to be very far from achieving the target of "harmonization" of practices. We have also demonstrated that there is a rather broad and heterogeneous bias induced in the measurement of several parameters by hemolysis and/or blood cell lysis. Basically, the rather heterogeneous and unpredictable response to hemolysis observed for several parameters both prevent the adoption of reliable statistical corrective measures of results on the basis of the degree of hemolysis (e.g., in the range of a free hemoglobin concentration between 0.3 and 1.3 g/L, the relative increase of measured potassium due to intracellular release varied broadly, from 0.7% to 3.0%). It is also worth mentioning that everyone would agree that the main bias might be roughly linearly dependent on the free hemoglobin concentration in the specimen and obviously independent from the initial concentration of the analyte, the mean variations observed for several parameters other than potassium (e.g., AST and LDH) exceed the quality specifications for desirable bias even in mild hemolytic specimens. Even more importantly, in the range of 1.3 to 10.3 g/L free hemoglobin (which includes the majority of specimens which are usually categorized as hemolyzed by visual inspection), the mean variation of the vast majority of clinical chemistry and coagulation parameters largely exceeds the quality specifications for desirable bias, thus preventing the identification of suitable ratios or corrective equations in correcting the values for the degree of hemolysis. The results of a previous investigation by Shepherd et al. are in keeping with ours [200]. Basically the bias between potassium in the hemolyzed and non-hemolyzed repeat samples was correlated with the serum HI measured on a Beckman Synchron LX20Pro. A significant linear correlation was proven, showing a potassium increase of approximately 0.16 mmol/L for each increment in the HI, Bland-Altman statistics in the patient-derived model showed the variability in predicted potassium concentration to be ±0.4 mmol/L. The magnitude of the relationship using authentic patient data with the Beckman LX20Pro HI was thereby too high to recommend the use of the HI to predict potassium concentration in hemolyzed specimens and obviate the need for repeat analysis; it can be used only to prioritize the need for urgent repeat testing. Finally, the use of these equations might only be suitable in the presence of spurious hyperkalemia due to in vitro hemolysis. But in the unfortunate circumstance that hemolysis has occurred in vivo, adjustment of the results is indeed misleading because it would not mirror a clinical situation that would otherwise require urgent therapeutic action (e.g., the timely recognition of the high risk associated with this electrolyte abnormality should stimulate measures to prevent deleterious side effects).

11.6.2 Report results with interpretative comments

An alternative approach to dealing with hemolyzed samples is to report data with an accompanying interpretation of the laboratory report, expressing either the potential range of the analyte concentration as predicted according to the degree of hemolysis (e.g., calculated from the HI) or simply stating that the result has been obtained on a hemolyzed specimen and, accordingly, might be biased owing to a certain degree of analytical and biological interference. In the latter case, results can be accompanied by some sort of alert or flag (e.g., "overestimation of K concentration: exclude in vivo hemolysis or repeat sampling") [201].

The concept of providing posttest guidance is widely utilized in laboratory practice [118]. It is also undeniable that this approach might be helpful to clinicians in the specific case of hemolysis (e.g., providing a rough estimation of the concentration of several analytes where hemolysis may be present), but it has also some clear limitations. In up to 97% to 98% of cases, hemolytic specimens stem from in vitro hemolysis, so that laboratory data obtained on these samples and transmitted to the wards would be biased and imprecise. There are instead firm recommendations (e.g., those issued by World Health Organization and the Italian Societies of Laboratory Medicine SIBioC and SIMeL) [202,203] that laboratory data obtained from unsuitable specimens should not be reported. It is also important to consider that an interpretive comment accompanying test results on the laboratory report (either paper or digital) might frequently go undetected or ignored, especially in the ED or ICU, where activities are more critical and frenzied than in other healthcare settings. In these circumstances, a direct notification (as detailed in Chapter 11.6.3) is indeed more effective for alerting the healthcare staff in charge of the patient. Finally, the laboratory data obtained on unsuitable specimens and therefore included within the laboratory report are permanently stored in the database of the Laboratory Information System (LIS), and this would represent unreliable and potentially misleading information in the longitudinal comparison of patients' data.

11.6.3 Suppress test results, warn clinicians, and ask for new specimens

Considering the limitations of the previously described strategies for the management of hemolyzed specimens, the suggested action is described in ▶Fig. 11.2. First it is strongly recommended to implement a policy of systematic inspection of the samples (when feasible – i.e., when there is no consolidation of preanalytical and analytical work stations) and eventual quantification of hemolysis (preferably using automatic estimation of the HI), especially when the degree of hemolysis is in the range that might be misinterpreted by simple visual inspection (i.e., from 0.3 to 0.9 g/L) and in those samples where hemolysis might be masked by an excess of other interfering substances, such as bilirubin and lipids. The use of delta checks, lack of "fit" with clinical details, implausible results, different results for the same analyte from different methods, and nonlinearity on dilution are additional indicators of the presence of a potential interfering substance in the sample, including free hemoglobin in serum or plasma (▶Tab. 11.6).

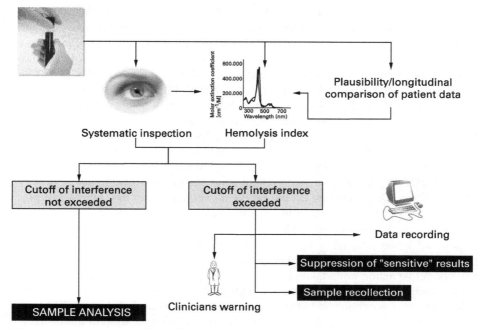

Fig. 11.2: Management of hemolyzed specimens.

Tab. 11.6: Management of specific laboratory tests (especially stat) according to the degree of hemolysis in the specimen

Degree of hemolysis	Free hemoglobin concentration	Test to be suppressed
Mild	≥0.3 to 0.6 g/L	• Potassium • Amino transferases • Lactate dehydrogenase
Frank	≥0.6 to 2.0 g/L	• Cardiac biomarkers • β–human chorionic gonadotropin • Glucose • Creatine kinase • Prothrombin time • Activated partial thromboplastin time • D-dimer
Gross	≥2.0 g/L	Virtually all tests

Further considerations are dependent upon the key concepts that very slight hemolysis has negligible (clinical) effects on most test values, modest to gross hemolysis might determine a dilutional effect on analytes present at a lower concentration in the RBCs as compared with plasma, while it might produce chemical/analytical interferences and spurious increase of analytes present at a higher concentration in RBCs than in plasma. Therefore when the degree of hemolysis reasonably exceeds the threshold of biological or analytical interference, the unsuitable sample should be systematically recorded. This process would make it possible to store and review data and then identify the preanalytical steps and clinical wards most susceptible to this type of error, thus providing an ideal basis for an efficient feedback to enable consideration of specific responsibilities. A hypothetical strategy based on the differential suppression of test results according to the HI is described in ▶Figs. 11.3 and ▶11.4, whereby results of samples with a HI ≤2 (i.e., free hemoglobin in serum or plasma ≤0.5 g/L) are reported, whereas results of "sensitive tests" such as potassium, AST, ALT, LDH, CK, iron and ammonia are suppressed in those samples with an HI between 2 and 7 (i.e., free hemoglobin in serum or plasma >0.5 and ≤3.0 g/L). All test results are suppressed in the presence of an HI >7 (i.e., free hemoglobin in serum or plasma >3.0 g/L). Moreover, an interpretive comment is systematically added to the laboratory report both for alerting clinicians on the problem encountered in the sample and providing a reliable explanation to justify the suppression of the tests, as follows: free hemoglobin in serum or plasma ≤0.5 g/L = no comment; free hemoglobin in serum or plasma >0.5 and ≤3.0 g/L = sample mildly hemolyzed; free hemoglobin in serum or plasma >3.0 g/L = sample frankly hemolyzed.

Any corrective action undertaken by the laboratory staff must also be recorded. The following step, especially when there is a high suspicion of hemolytic anemia, is the establishment of a good liaison between the laboratory staff and clinicians. This is of paramount importance for rapid notification and reliable identification of the underlying cause, frequently a life-threatening disorder. Therefore the clinical wards should be immediately alerted, so that in vivo hemolysis can be safely ruled out. Concomitantly,

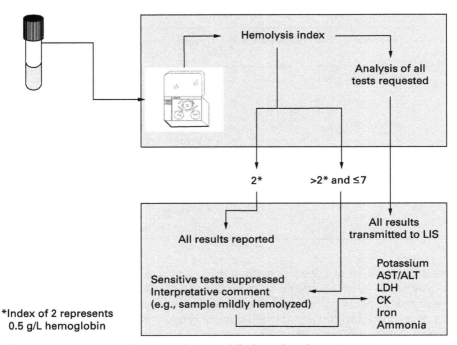

Fig. 11.3: Management of test results on mildly hemolyzed specimens.

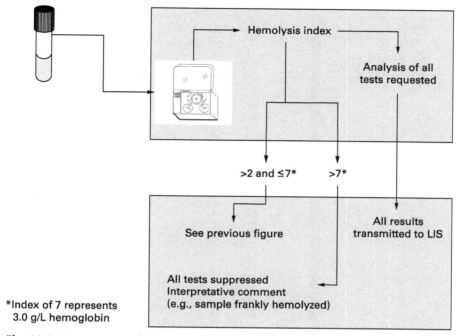

Fig. 11.4: Management of test results on frankly hemolyzed specimens.

all tests influenced by the presence and/or extent of hemolysis should be suppressed (according to local policies for instrumentation and methods), and the laboratory staff should request a recollection of the specimen (▶Fig. 11.2). The practice of asking for a new sample is further profitable, since it allows for confirmation or refutation of the hypothesis that the patient might be suffering from intravascular hemolysis, depending on whether the second sample is also hemolytic and shows a comparable HI.

This approach might, however, carry a "relational" problem with the stakeholders of laboratory testing – that is, clinicians (especially emergency physicians or those working in ICUs) and patients. Salinas et al. reported that using the analyte-specific HI (Modular P800, Roche Diagnostics), the number of laboratory test results that would be suppressed in hemolyzed specimens is as high as 8.3% for LDH (suppressed in samples with HI >20), 5.6% for AST (suppressed in samples with HI >25), 2.7% for total bilirubin (suppressed in samples with HI >50), 2.4% for potassium (suppressed in samples with HI >50), 2.2% for iron (suppressed in samples with HI >50), 1.7% for ALT (suppressed in samples with HI >60), 0.6% for CK (suppressed in samples with HI >100) and 0.2% for GGT (suppressed for samples with HI >200) [204]. The estimated number of patients who would have their blood recollected according to these figures is about 0.2%. Thus the burden of organizational (recollection), clinical (repeated venipuncture), and relational (i.e., dispute with clinicians for not reporting test results) issues is overall expected to be quite manageable.

12 The challenge of synthetic blood substitutes

Blood substitutes have long been sought for replacement of "bad humor," treatment of chronic and acute anemias, and rapid replacement of blood lost after trauma [205]. The last indication has understandably stimulated a great deal of interest from the military services. Therefore there is a long history in science of the attempt to develop artificial substitutes for damaged body parts. Although some body parts (e.g., teeth and limbs) might be reliably replaced by a kind of "imitation" without major loss of functionality, the development of a substitute for RBCs has proven at least partially elusive so far. Historically, blood loss from critical surgery or trauma has been dealt with either through the use of volume-replacing fluids or transfusions. Remarkable advances in chemical and biotechnological research has, however, allowed the development of a novel approach to massive blood losses represented by temporary oxygen carriers, widely known as "blood substitutes." The blood substitutes currently available are chiefly polymerized hemoglobin solutions, hemoglobin-based oxygen carriers (HBOCs), perfluorocarbons (PFCs) [206], and RBC-mimicking synthetic biomaterial particles [207,208]. The leading clinical advantages of these blood substitutes include larger supply and availability, lower risk of the transmission of blood-borne pathogens, room-temperature storage, and extended survival of stored materials, longer shelf life, absence of RBC antigens, and no risk of immune incompatibility using group O RhD-negative RBCs.

Nevertheless, these new products represent a major challenge to clinical laboratories. Since they are intensely colored and used in high concentrations therapeutically, it is likely that such products would generate significant analytical interference with the current technology used in most laboratories. Moreover, much testing on the blood of patients receiving these products will be impossible to perform reliably [209,210]. Moreover, the simple translation of interference studies carried out on hemolyzed specimens to evaluate the influence of blood substitutes on laboratory testing is unreliable, inasmuch as the hemolyzed serum or plasma specimens contain a variety of intracellular contaminants and cellular debris released from damaged blood cells as well as cells (e.g., a modest number of platelets), antibodies, coagulation factors, and a variety of blood components. All these materials are instead generally lacking in HBOC or PFC solutions, which instead only contain one potentially interfering substance (i.e., HBOC or PFC). Therefore when the bias is caused only by the presence of a blood substitute in the specimen and not by a generalized process of blood cells lysis, the use of corrective formulas developed in the latter setting is unreliable, since it would not consider the contribution of intracellular components and stroma.

12.1 Individual studies evaluating the interference of blood substitutes

Ma et al. examined the effects of a stock bovine 130 g/L HBOC preparation and a second-generation (PFC) emulsion (Perflubron®) on a vast array of clinical chemistry, immunoassay, therapeutic drug, and coagulation tests as measured on the Vitros 750, Hitachi 747, Abbott IMx, and AxSym as well as the BBL Fibro-System and MLA 1000 C analyzers [210]. A significant negative interference by HBOC was observed on calcium, phosphorus, amylase, AST, magnesium, total bilirubin, and conjugated bilirubin on the Vitros 750. On the Hitachi 747, a significant positive interference was observed on CO_2, creatinine (colorimetric Jaffe method), albumin, calcium, AST, cholesterol and uric acid. A negative interference was observed on ALT and lipase on the Vitros 750, as well as on ALP, ALT, total bilirubin, GGT, and LDH on the Hitachi 747. Although no significant interference from HBOC was observed in several immunoassays (i.e., troponin I, thyroid-stimulating hormone, digoxin, phenytoin, lidocaine, N-acetylprocainamide, procainamide, quinidine, and theophylline), CK-MB and vancomycin were negatively biased, whereas gentamycin exhibited a positive bias. The interference of HBOC in coagulation testing was basically method-dependent. Although PT, APTT, and fibrinogen using the BBL Fibro-System, which utilizes electrical clot detection, were unaffected at HBOC concentrations up to 50 g/L, all these clotting test produced a "no clot detected" error flag at 20 g/L HBOC on the MLA 1000C analyser, which detects clot formation optically. Finally, manual latex agglutination assays for D-dimer and fibrin degradation products (FDPs) were unaffected at up to 50 g/L HBOC. At variance with HBOC, the PFC-based oxygen carrier (600 g/L) caused no interference on any of these chemistry or immunoassay tests except for amylase (negative bias) and ammonia (positive bias) on the Vitros 750 and plasma iron (negative bias) on the Hitachi 747. It was thus concluded that neither bovine Hb-based HBOC at plasma hemoglobin concentrations up to 50 g/L nor Perflubron® emulsion at maximal dosing rates interfere in critical care assays that might be urgently requested perioperatively.

In a further investigation, Moreira et al. investigated the effect of an ultrapurified HBOC solution (Hemopure®) on the performance of Ektachem and Hitachi clinical analyzers [211]. The instruments of nine hospitals participating in the Hemopure® clinical trials were challenged with samples containing Hemopure® up to 50 g/L to determine the maximal amount of HBOC that may be present in a sample to obtain accurate results for 22 clinical chemistry tests. Instrument results were compared with the theoretical value and considered free from interference if percent differences did not exceed 5% for electrolytes, 20% for enzymes, and 10% for all other parameters. Sodium, potassium, chloride, AST, calcium, urea nitrogen, CK, and glucose were unaffected by concentrations of HBOC as high as 50 g/L HBOC. Conversely, both instruments exhibited significant biases in the measurement of albumin (positive), bilirubin (positive), GGT (positive bias or unable to quantify), ALP (unable to quantify), and LDH (negative bias or unable to quantify). Other parameters showed various degrees of interference (e.g., amylase-positive bias, ALT-positive bias, and total protein–positive bias, or unable to quantify on Ektachem; cholesterol-positive bias on Hitachi).

Ali and Campbell investigated the effects of an HBOC (i.e., o-raffinose cross-linked hemoglobin, Hemolink®) on selected routine clinical chemistry tests assayed on a Hitachi 717 with reagents purchased from Boehringer Mannheim [209]. The HBOC Hemolink® is prepared from hemoglobin purified from outdated human RBCs. The purified hemoglobin reacts with o-raffinose, generating two hemoglobin {alpha}ß dimers covalently cross-linked between amino groups within the 2,3-diphosphoglycerate binding pocket to form stable hemoglobin tetramers of 64 kDa. The o-raffinose also interacts with surface amino groups to form intermolecular linkages, producing stable hemoglobin polymers (128 to 600 kDa). The final Hemolink® preparation contains about 40% of the hemoglobin tetramer and some 60% of the hemoglobin polymers. The study design was based on spiking three normal plasma pools and one "pathological" plasma pool (i.e., characterized by increased analyte concentrations) with two lots of Hemolink® to obtain final concentrations of 0.0 (control), 1.25, 2.5, 5.0, 7.5, 10, 15, and 20 g/L, measured as total hemoglobin. Interference was calculated in terms of the absolute or relative error, and the interference was considered to achieve significance when the relative error exceeded a limit of 3 × CV of each method. Moreover, when the observed interference did not exceed the combined analytical and biological variability, Hemolink® was not considered to interfere with the clinical interpretation of test results. The data published in the article attest that no interference could be observed for sodium, potassium, chloride, urea, glucose, total protein, CK, AST, calcium, or phosphorus up to 20 g/L HBOC. Nevertheless a clinically significant bias was recorded for albumin and ALT at Hemolink® concentrations >5.0 g/L; for creatinine, total bilirubin, LDH, and GGT at Hemolink® concentrations >1.25 g/L and for ALP and iron at all Hemolink® concentrations tested. No significant differences were observed in the absolute or relative error between the normal and abnormal sample pools tested or between the two lots of Hemolink® tested. It is noteworthy that the interference in test results was directly proportional to the concentration of HBOC in the specimens and was independent of the concentration of the analytes in the range of values tested except for bilirubin. It is also interesting that the bias in grams per liter with Hemolink® was significantly different from that of hemolysis on these tests, since the presence of intracellular contaminants in hemolyzed specimens released from damaged bold cells also contributes to the overall interference in testing some analytes such as LDH, AST, CK and potassium.

In a further study, the same group of authors assessed the interferences of Hemolink® on three serum creatinine assays based on different principles (i.e., a Boehringer Mannheim kinetic Jaffe method based on a two-reagent system for the Hitachi 717, a Synermed kinetic Jaffe method based on a two-reagent system for the Hitachi 717, and a Johnson & Johnson Clinical Diagnostics kinetic enzymatic assay in which the rate of formation of the final oxidized leuco dye is monitored at 670 nm on the Vitros 750) [212]. The HBOC Hemolink® was prepared as a 100 g/L solution (measured as total hemoglobin) and added to serum pools containing creatinine concentrations within the reference interval to obtain approximate Hemolink® concentrations of 0.0 (control sample), 1.25, 2.5, 5.0, 7.5, 10, 15, and 20 g/L. Moreover, normal (i.e., creatinine concentration from 44 to 97 µmol/L) and abnormal serum pools (i.e., creatinine concentration from 147 to 548 µmol/L) were also diluted 1:1 with various amounts of Hemolink® to obtain approximate Hemolink® concentrations of 0.0 (control sample), 10, 20, 30, 40, and 50 g/L. Two lots of Hemolink® were tested to assess lot-to-lot consistency of the bias. Absolute and relative errors were plotted as a function of the Hemolink® concentration in each sample and the interference was deemed to be analytically significant when it exceeded the 95% confidence limits for each method. The results of the study showed that Hemolink® exerted a positive bias with the Boehringer Mannheim assay on the Hitachi 717, since recovery of creatinine in samples containing 20 g/L HBOC increased to 224% ± 14% of control values in samples with normal creatinine values and to 143% ± 7% of control values in samples with increased creatinine. The bias on the Boehringer Mannheim/Hitachi 717 assay exceeded the 95% confidence limits for Hemolink® concentrations >5.0 g/L, and the absolute error in the measurement showed a linear relationship with the HBOC concentration (i.e., the error was 5.4 μmol/L creatinine per 1 g/L Hemolink® in the normal samples, and 7.9 μmol/L creatinine per 1 g/L Hemolink® in abnormal samples). Conversely, both the Synermed/Hitachi 717 method and the Ortho/Vitros 750 method showed no significant interference from Hemolink at concentrations as high as 20 g/L in the initial studies. Moreover, all results for normal and abnormal pools supplemented with 1.25 to 15 g/L of Hemolink® were within the 95% confidence limits of the assay. However, the Synermed assay overestimated creatinine concentrations at the lower end of the reference range when Hemolink® concentrations exceeded 30 g/L, even though this bias was not considered to be clinically significant.

Kazmierczak et al. investigated the effects of diaspirin cross-linked hemoglobin chains (DCLHb; also known as HemAssist; Baxter Healthcare Corp., Deerfield, IL) on total calcium, cholesterol, total bilirubin, and potassium using a Hitachi 747 [213]. For each analyte, two pools of serum were initially prepared, the former containing the analyte of interest at high concentration and the latter containing the analyte at low concentration. The analyte concentrations in the low and high serum pools used to make up the linear series of specimen pools included, at a minimum, the 1% and 99% percentiles of patient results. Four pools containing intermediate concentrations of the analyte were also prepared by scalar mixing of the low and high pools, which were further divided into aliquots in which DCLHb was added to achieve final hemoglobin concentrations of 0.0, 2.5, 5.0, 7.5, 10.0, 15.0, and 20.0 g/L. The limits for establishing the presence of clinically significant interference were those specified in the Clinical Laboratory Improvement Amendments of 1988 (CLIA-88). Basically, DCLHb interfered with the measurement of calcium, cholesterol, and total bilirubin but did not interfere with that of potassium. Although statistically significant interference by DCLHb was noted for calcium, the magnitude of interference did not achieve clinical significance. The measurement of cholesterol and total bilirubin showed instead significant statistical and clinical interference caused by the presence of DCLHb.

Chance et al. assessed the interference of a commercially developed HBOC with the assay of alkaline phosphatase as measured on a Hitachi 917 analyzer and ultraviolet-visible spectrophotometry [214]. The polymerized hemoglobin (PolyHeme; Northfield Laboratories Mount Prospect, Illinois, USA) consists of native tetrameric human hemoglobin polymerized using glutaraldehyde and modified by pyridoxal phosphate to optimize the P50 value; it is free of all unreacted tetramer and has a hemoglobin concentration of 100 g/L. In the experimental conditions, PolyHeme displayed substantial absorbance at 415 nm, the wavelength of analysis used to measure the formation of 4-nitrophenol, which produced a strong negative interference plot (i.e., constant decrease of approximately −50 U/L per 10 g/L HBOC) because of alkali denaturation of the substitute. An identical interference was observed for native hemoglobin (hemolysate), indicating that the interference was not derived from the polymerization process. The interference can be corrected by implementing a rate-correction procedure, or it can be avoided by measurement at 450 nm.

Most recently Cameron et al. assessed the interference of the HBOC Hemospan® (i.e., a solution of polyethylene glycol–conjugated human hemoglobin in development for use in volume expansion during elective surgical procedures) on 35 clinical laboratory tests [215]. Starting with 2.0 mL of deidentified patient serum pools, increasing volumes of the Hemospan® stock solution were added to obtain final HBOC concentrations of 0.0 g/L, 5.0 g/L, 10.0 g/L, 15.0 g/L, and 20.0 g/L. A bias exceeding deviations of

±10% in the Hemospan®-spiked sample was reported as "negative" or "positive" interference. No Hemospan® interference was noticed for the measurement of sodium, potassium, chloride, calcium, lactate, osmolality, total CK, CK-MB, ALT, bilirubin (direct and total), blood urea nitrogen, CO_2, C-reactive protein, glucose, HDL cholesterol, LDL-cholesterol, lipase, or transferrin. Conversely, the HBOC caused a negative bias in the measurement of serum creatinine (enzymatic method), amylase, ALP, uric acid, and GGT assays as well as a positive interference in serum phosphate, LDH, iron, triglycerides, total protein, AST, cholesterol, magnesium, and albumin tests. Moreover, a positive bias was observed on the measurement of troponin I. This suggests that this HBOC interferes not only in the measurement of some analytes by spectral or biophysical properties but also with troponin, other sample components, or the assay reagent to alter the integrity of this immunoassay.

Ali et al. investigated the effects of five hemoglobin-based blood substitutes on the measurements of eight different oximeters and cooximeters (i.e., AVL Omni 6, AVOXimeters 1000 and 4000, Ciba Corning CC270 CO-Oximeter, Instrumentation Laboratory Synthesis 35, IL482 and IL682 CO-Oximeters, and Radiometer OSM3 Hemoximeter) [216]. The five blood substitutes were obtained from Apex Bioscience (human hyridoxalated hemoglobin polyoxyethylene), Baxter Healthcare Corp. (human dibromosalicylate bisfumarate cross-linked alpha-chains), Biopure Corp. (bovine glutaraldehyde polymerized hemoglobin), Hemoglobin Therapeutics (recombinant human covalent alpha-chain dimers of Hb Presbyterian) and Hemosol, Inc. (human o-raffinose cross-linked polyhemoglobin). Globally, the instruments yielded measurements of total hemoglobin concentration in undiluted blood substitutes as variable as those obtained on the control material. Conversely, when compared with readings on controls, the test instruments yielded measurements of the fractional concentrations of oxy-, deoxy-, carboxy-, and methemoglobin that showed greater instrument-to-instrument disparities and larger standard deviations about the all-instrument means. In some cases, the interference was even more obvious: five of six cooximeters gave negative carboxyhemoglobin readings on one particular product. However, taken together, these results indicate that the instruments tested produced slightly less accurate but still clinically useful measurements in the presence of hemoglobin-based blood substitutes.

Jahr et al. investigated whether three HBOCs (i.e., hemoglobin glutamer-200 [bovine] HBOC-200, Oxyglobin; hemoglobin glutamer-250 [bovine] HBOC-201, Hemopure; and hemoglobin raffimer, Hemolink®) interfered with the accuracy of lactate measurements [217]. Combinations of concentrated L-lactate solution, HBOC, and blood or plasma were added to sample tubes to make a linear and constant increase in lactate concentration (from 5 to 110 mg/dL) in consecutive samples. Comparisons were made between instrument-assayed and calculated lactate concentrations. For Hb glutamer-250, the average difference between measured and calculated lactate concentrations was −5.1 mg/dL, with greater underestimation at larger lactate concentrations. For Hb raffimer, the average difference was −2.2 mg/dL. The Hb glutamer-200, which was tested on three different analyzers, produced differences ranging between −8.4 mg/dL and +1.3 mg/dL. It was therefore concluded that increasing levels of an HBOC in plasma determine inaccurate values of lactate (especially at higher lactate concentrations), causing underestimation of measured values and potential undertreatment of the patient.

The reliability of lactate measurements in the presence of methemoglobin from an oxidized bag of the HBOC hemoglobin glutamer-200 (Oxyglobin) was investigated by Osgood et al. [218]. The mean difference between measured and calculated lactate was −5.1 mg/dL (1% Oxyglobin), −5.8 mg/dL (10% Oxyglobin), −4.6 mg (20% Oxyglobin), and −8.5 mg/dL (40% Oxyglobin). These results confirm that true lactate levels in the presence of Oxyglobin are underestimated when measured, independent of the amount of Oxyglobin present in the specimen.

Moon-Massat et al. determined the feasibility of laboratories to use an instrument's HI to verify the accuracy of test results in the presence of Hemopure (HBOC-201), which is the only HBOC commercially licensed for treatment of acute anemias in humans (HBOC-201 contains 130 g/L soluble polymerized bovine hemoglobin) [219]. The potential interference of Hemopure on 24 common laboratory teats was also evaluated by using two commercial quality control materials and two plasma pools with normal and abnormal concentration of the analytes. For each of the four samples tested, a series of aliquots was prepared containing from 0.0 to 65 g/L of Hemopure in increments of 5 g/L. RBC lysate hemoglobin was also spiked to serum and the instrument manufacturer's HI limits as measured on Roche Hitachi Modular P800 were compared with HI limits for reporting the same laboratory assays determined by adding Hemopure. The criteria for determining at what concentration of HBOC the results of the single tests were interfered with were based upon either CLIA or College of American Pathology limits. Notably, regression analysis of the mean HI values for all samples at each prepared Hemopure concentration revealed an almost perfect linear relationship ($y = 0.88x − 0.02$, $r^2 = 0.99$), regardless of the sample matrix. Twenty-one analytes showed Hemopure limits almost identical to or even more conservative than those obtained from the HBOC-201 limit determined by direct interference testing and/or the manufacturer's HI limits. For these analytes, therefore, either the hemolysis interference data or the manufacturer's HI thresholds might be used to make decisions as to whether or not test results in the presence of HBOCs should be reported. Conversely, creatinine and uric acid (both positively biased on the Trinder reaction method, which is based on the conversion of aminophenazone to a quinine imine dye by peroxidase), as well as albumin showed a more conservative limit when direct Hemopure interference testing was assayed (calculated HI limits were 0 to 5 for both uric acid and albumin vs. 0 for creatinine, vs. manufacturer's HI >10 for all). As regards creatinine and uric acid, the authors hypothesized that Hemopure might posses a greater peroxidase activity than endogenous hemoglobin (i.e., chemical interference), likely attributable to its polymerized nature, which would amplify the production of the quinine imine dye. On the other hand, the underlying mechanism or mechanisms responsible for the bias in albumin measurement are still unclear, even though this may be attributable to the fact that polymerized HBOC is more stable at reduced pH and might thereby exert a larger effect at the reaction pH of 4.1. Nevertheless, the leading evidence coming from this study is that the HI can be used to approximate a sample's HBOC concentration without the use of a direct hemoglobin measurement.

12.2 Detection and quantification of blood substitutes

Routine monitoring of circulating cell-free hemoglobin is useful for evaluating the efficacy of blood substitute administration, assessing the clearance from circulation, and obtaining reliable hematological test results (e.g., RBC indices, MCH, and MCHC) in patients undergoing therapy with blood substitutes. The colorimetric methods typically used to assess the hemoglobin (e.g., Drabkin's assay) cannot, however, differentiate between cell-associated and cell-free hemoglobin and are therefore unsuitable for measuring the concentration of hemoglobin substitutes in blood. To fill this gap, Kunicka et al. developed a combined cell-based assay and colorimetric determination that accurately discriminate between cell-associated and cell-free hemoglobin over a wide range of hemoglobin levels [220]. Specifically, the methods provide simultaneous measures of RBC volume and hemoglobin concentration of individual RBCs, so that it is possible to obtain an independent measure of cell-associated hemoglobin as follows: RBC × MCV × CHCM. This value of cell-associated hemoglobin is then subtracted from the total hemoglobin content of the blood, as measured colorimetrically following lysis of RBCs by a standard hemoglobin reagent, to provide an accurate estimation of cell-free hemoglobin. The method has already been shown to be reliable for discriminating between cellular and cell-free hemoglobin for the two blood substitutes Hemopure and Oxyglobin over a wide range of hemoglobin values (i.e., from 0 to 130 g/L).

Although blood substitutes have been principally developed for the treatment of acute and life-threatening anemias (i.e., acute blood loss due to trauma or surgery, acute blood loss in Jehovah's Witnesses, or in cases of multiple RBC alloantibodies, rare blood type, endemic infection in donor, extender in acute normovolemic hemodilution, nitric oxide scavenger in septic shock, anti-ischemic therapy in sickle cell crisis, percutaneous transluminal coronary angioplasty, myocardial infarction, cardiopulmonary bypass, vaso-occlusive stroke, ex vivo organ or tissue preservation, neuroprotectant in cardiopulmonary bypass, sensitizer for chemotherapy and radiotherapy, partial liquid ventilation for acute respiratory distress syndrome, near drowning, smoke inhalation, and infections) [221], a sinister scenario is emerging, which is their translation as unfair practices in sports, in analogy with what already happened for several other forms of blood boosting. Therefore HBOCs have been included in the International Olympic Committee and World Anti-Doping Agency (WADA) lists of substances and methods prohibited in sports and a reliable screening assay has been developed to identify their unfair usage [222]. The method is based on electrophoresis of serum samples cleared of haptoglobin. Four successive steps (immunoprecipitation of haptoglobin, electrophoresis of the cleared serum, Western blotting of the separated proteins, and detection of hemoglobin-related molecules based on the peroxidase properties of the heme moiety) can provide electropherograms that can easily be interpreted in terms of the presence of HBOCs. This method was tested and judged to be reliable (e.g., the test was able to detect Hemopure for 4 to 5 days after administration of 45 g to healthy individuals) and accurate with various types of HBOCs (i.e., polymerized, conjugated, and cross-linked hemoglobins). Even more importantly, the test demonstrated no possible confusion with endogenous hemoglobin that may be present in cases of hemolysis. An additional approach has been proposed by Varlet-Marie et al. [223]. This alternative method is based on a size-exclusion HPLC technique. The chromatograms of all HBOCs tested were clearly separated from the peak attributable to human hemoglobin dimers, and it was also possible to discriminate the different HBOCs (when products were present in a high concentration) on the basis of their specific chromatographic profiles. Moreover, the differences were distinguishable on the basis of either presence/absence of the peaks or the separation between respective peaks. The profiles for human serum samples collected immediately after the infusion of Hemopure also showed a distinctive profile, which remained consistent for at least 48 hours. Further methods were also developed, based on such techniques as electrophoretic size-exclusion HPLC, liquid chromatography-electrospray ionization ion trap mass spectrometric detection [224], and direct visual screening of plasma discoloration [225], even though those previously described by Lasne et al. and Varlet-Marie et al. appear to be the most accurate, reliable, and suitable for antidoping campaigns.

12.3 Final remarks on blood substitutes

Although available studies in the scientific literature will provide useful guidelines, individual laboratories should assess the interferences of blood substitutes (especially HBOC) in light of the myriad testing methods in use, differences in the various products, and institution-specific opinions as to what constitutes significant interference.

References

1. Franco RS. The measurement and importance of red cell survival. Am J Hematol 2009;84:109–114.
2. Palis J, Segel GB. Developmental biology of erythropoiesis. Blood Rev 1998;12:106–114.
3. Lippi G, Blanckaert N, Bonini P, Green S, Kitchen S, Palicka V, et al. Haemolysis: an overview of the leading cause of unsuitable specimens in clinical laboratories. Clin Chem Lab Med 2008;46:764–772.
4. Lippi G, Plebani M, Di Somma S, Cervellin G. Hemolyzed specimens: a major challenge for emergency departments and clinical laboratories. Crit Rev Clin Lab Sci 2011;48:143–153.
5. Guder W. Haemolysis as an influence and interference factor in clinical chemistry. J Clin Chem Clin Biochem 1986;24:125–126.
6. Kurec A, Wyche KL. Institute for Quality in Laboratory Medicine Series: controversies in laboratory medicine: nursing and the laboratory: relationship issues that affect quality care. MedGenMed 2006;8(3).
7. Lippi G, Guidi GC, Mattiuzzi C, Plebani M. Preanalytical variability: the dark side of the moon in laboratory testing. Clin Chem Lab Med 2006;44:358–365.
8. Lippi G, Guidi GC. Risk management in the preanalytical phase of laboratory testing. Clin Chem Lab Med 2007;45:720–727.
9. Jones BA, Calam RR, Howanitz PJ. Chemistry specimen acceptability: a College of American Pathologists Q-Probes study of 453 laboratories. Arch Pathol Lab Med 1997;121:19–26.
10. Plebani M, Ceriotti F, Messeri G, Ottomano C, Pansini N, Bonini P. Laboratory network of excellence: enhancing patient safety and service effectiveness. Clin Chem Lab Med 2006;44:150–160.
11. Gonzalez-Porras JR, Graciani IF, Alvarez M, Pinto J, Conde MP, Nieto MJ, et al. Tubes for pretransfusion testing should be collected by blood bank staff and labeled until the implementation of new technology for improved sample labelling. Results of a prospective study. Vox sanguinis 2008;95:52–56.
12. Carraro P, Servidio G, Plebani M. Hemolyzed specimens: a reason for rejection or a clinical challenge? Clin Chem 2000;46:306–307.
13. Romero A, Muñoz M, Ramos JR, Campos A, Ramírez G. Identification of preanalytical mistakes in the stat section of the clinical laboratory. Clin Chem Lab Med 2005;43:974–975.
14. Burns ER, Yoshikawa N. Hemolysis in serum samples drawn by emergency department personnel versus laboratory phlebotomists. Lab Med 2002;33:378–380.
15. Lippi G, Salvagno GL, Favaloro EJ, Guidi GC. Survey on the prevalence of hemolytic specimens in an academic hospital according to collection facility: opportunities for quality improvement. Clin Chem Lab Med 2009;47:616–618.
16. Salvagno GL, Lippi G, Bassi A, Poli G, Guidi GC. Prevalence and type of pre-analytical problems for inpatients samples in coagulation laboratory. J Eval Clin Pract 2008;14:351–353.
17. European Preanalytical Scientific Committee. Available at: http://specimencare.com.
18. Lippi G, Ippolito L, Fontana R. Prevalence of hemolytic specimens referred for arterial blood gas analysis. Clin Chem Lab Med 2011;49:931–932.

19. Hashimoto C. Autoimmune hemolytic anemia. Clin Rev Allergy Immunol 1998;16: 285–295.
20. Bessler M, Schaefer A, Keller P. Paroxysmal nocturnal hemoglobinuria: In sights from recent advances in molecular biology. Transfus Med Rev 2001;15:255–267.
21. Chiao EY, Engels EA, Kramer JR, et al. Risk of immune thrombocytopenic purpura and autoimmune hemolytic anemia among 120 908 US veterans with hepatitis C virus infection. Arch Intern Med 2009;169:357–363.
22. Gehrs BC, Friedberg RC. Autoimmune hemolytic anemia. Am J Hematol 2002;69:258–271.
23. Hashimoto C. Autoimmune hemolytic anemia. Clin Rev Allergy Immunol 1998;16: 285–295.
24. Packman CH. Hemolytic anemia due to warm autoantibodies. Blood Rev 2008;22: 17–31.
25. Tabbara IA. Hemolytic anemias. Diagnosis and management. Med Clin North Am 1992;76:649–668.
26. Packman CH, Leddy JP. Acquired hemolytic anemia due to warm-reacting autoantibodies. In: Beutler E, Lichtman MA, Coller BS, Kipps TJ, eds. Williams Hematology. 5th ed. New York: McGraw Hill; 1995:667–684.
27. Packman CH, Leddy JP. Cryopathic hemolytic syndromes. In: Beutler E, Lichtman MA, Coller BS, Kipps TJ, eds. Williams Hematology. 5th ed. New York: McGraw Hill; 1995:685–690.
28. Beutler E. Hemolytic anemia due to infections with microorganisms. In: Beutler E, Lichtman MA, Coller BS, Kipps TJ, eds. Williams Hematology. 5th ed. New York: McGraw Hill; 1995:674–676.
29. Packman CH, Leddy JP. Drug-related immune hemolytic anemia. In: Beutler E, Lichtman MA, Coller BS, Kipps TJ, eds. Williams Hematology. 5th ed. New York: McGraw Hill; 1995:691–696.
30. Beutler E. Hemolytic anemia due to chemical and physical agents. In: Beutler E, Lichtman MA, Coller BS, Kipps TJ, eds. Williams Hematology. 5th ed. New York: McGraw Hill; 1995:670–673.
31. Wright MS. Drug-induced hemolytic anemias: Increasing complications to therapeutic interventions. Clin Lab Sci 1999;12:115–118.
32. Martinez J. Microangiopathic hemolytic anemia. In: Beutler E, Lichtman MA, Coller BS, Kipps TJ, eds. Williams Hematology. 5th ed. New York: McGraw Hill; 1995:669.
33. George JN. The thrombotic thrombocytopenic purpura and hemolytic uremic syndromes: overview of pathogenesis (Experience of The Oklahoma TTP-HUS Registry, 1989–2007). Kidney Int Suppl 2009:S8–S10.
34. Mecozzi G, Milano AD, De Carlo M, et al. Intravascular hemolysis in patients with new-generation prosthetic heart valves: A prospective study. J Thorac Cardiovasc Surg 2002;123:550–556.
35. Modell B, Darlison M. Global epidemiology of haemoglobin disorders and derived service indicators. Bull World Health Org 2008;86:480–487.
36. Frenette PS. Sickle cell vasoocclusion: a multistep and multicellular paradigm. Curr Opin Hematol 2002;9:101–106.
37. Beutler E. G6PD: population genetics and clinical manifestations. Blood Rev 1996;10: 45–52.
38. Mehta A, Mason PJ, Vulliamy TJ. Glucose-6-phosphatase dehydrogenase deficiency. Baillieres Best Pract Clin Haematol 2000;13:21–38.
39. Telford RD, Sly GJ, Hahn AG, Cunningham RB, Bryant C, Smith JA. Footstrike is the major cause of hemolysis during running. J Appl Physiol 2003;94:38–42.
40. Smith JA. Exercise, training and red blood cell turnover. Sports Med 1995;19:9–31.

41. Tobal D, Olascoaga A, Moreira G, Kurdián M, Sanchez F, Roselló M, Alallón W, Martinez FG, Noboa O. Rust urine after intense hand drumming is caused by extracorpuscular hemolysis. Clin J Am Soc Nephrol 2008;3:1022–1027.
42. Córdova Martínez A, Villa G, Aguiló A, Tur JA, Pons A. Hand strike-induced hemolysis and adaptations in iron metabolism in Basque ball players. Ann Nutr Metab 2006;50: 206–213.
43. Weitz IC. Thrombosis in patients with paroxysmal nocturnal hemoglobinuria. Semin Thromb Hemost 2011;37(3):315–321.
44. Favaloro EJ, Lippi G, Adcock DM. Preanalytical and postanalytical variables: the leading causes of diagnostic error in hemostasis? Semin Thromb Hemost 2008;34:612–634.
45. Lippi G. Governance of preanalytical variability: travelling the right path to the bright side of the moon? Clin Chim Acta 2009;404:32–36.
46. Lippi G, Chance JJ, Church S, Dazzi P, Fontana R, Giavarina D, et al. Preanalytical quality improvement: from dream to reality. Clin Chem Lab Med 2011;49:1113–1126.
47. Ogiso T, Iwaki M, Yamamoto M. Hemolysis induced by benzyl alcohol and effect of the alcohol on erythrocyte membrane. Chem Pharm Bull (Tokyo) 1983;31:2404–2415.
48. Tamechika Y, Iwatani Y, Tohyama K, Ichihara K. Insufficient filling of vacuum tubes as a cause of microhemolysis and elevated serum lactate dehydrogenase levels. Use of a data-mining technique in evaluation of questionable laboratory test results. Clin Chem Lab Med 2006;44:657–661.
49. Lippi G, Salvagno GL, Montagnana M, Franchini M, Guidi GC. Venous stasis and routine hematologic testing. Clin Lab Haematol 2006;28:332–337.
50. Lippi G, Salvagno GL, Montagnana M, Brocco G, Guidi GC. Influence of short-term venous stasis on clinical chemistry testing. Clin Chem Lab Med 2005;43:869–875.
51. Lippi G, Salvagno GL, Montagnana M, Guidi GC. Short-term venous stasis influences routine coagulation testing. Blood Coagul Fibrinolysis 2005;16:453–458.
52. Rosenfled G, Schenberg S, Nahas L. Fluid absorption by red blood cells and hemolysis in experimental venous stasis. Mem Inst Butantan 1957–1958;28:237–244.
53. Saleem S, Mani V, Chadwick MA, Creanor S, Ayling RM. A prospective study of causes of haemolysis during venepuncture: tourniquet time should be kept to a minimum. Ann Clin Biochem 2009;46(Pt 3):244–246.
54. Serdar MA, Kenar L, Hasimi A, Koçu L, Türkmen YH, Kurt I, et al. Tourniquet application time during phlebotomy and the influence on clinical chemistry testing; is it negligible? Turk J Biochem 2008;33:85–88.
55. Stankovic AK, Smith S. Elevated serum potassium values: the role of preanalytic variables. Am J Clin Pathol 2004;121(Suppl):S105–S112.
56. Becton Dickinson White Paper VS5391: Evaluation of Sample Quality and Analytic Results between Specimens Collected in Vacutainer Tubes and Current Syringe Collections. Franklin Lakes, NJ: Becton Dickinson; 2001.
57. Grant MS. The effect of blood drawing techniques and equipment on the hemolysis of ED laboratory blood samples. J Emerg Nurs 2003;29:116–121.
58. Wilcox GJ, Barnes A, Modanlou H. Does transfusion using a syringe infusion pump and small-gauge needle cause hemolysis? Transfusion 1981;21(6):750–751.
59. Sharp MK, Mohammad SF. Scaling of hemolysis in needles and catheters. Ann Biomed Eng 1998;26:788–797.
60. Kennedy C, Angermuller S, King R, Noviello S, Walker J, Warden J, et al. A comparison of hemolysis rates using intravenous catheters versus venipuncture tubes for obtaining blood samples. J Emerg Nurs 1996;22:566–569.
61. Halm MA, Gleaves M. Obtaining blood samples from peripheral intravenous catheters: best practice? Am J Crit Care 2009;18:474–478.

62. Ong ME, Chan YH, Lim CS. Reducing blood sample hemolysis at a tertiary hospital emergency department. Am J Med 2009;122:1054.e1–6.
63. Lowe G, Stike R, Pollack M, Bosley J, O'Brien P, Hake A, et al. Nursing blood specimen collection techniques and hemolysis rates in an emergency department: analysis of venipuncture versus intravenous catheter collection techniques. J Emerg Nurs 2008;34:26–32.
64. Dugan L, Leech L, Speroni KG, Corriher J. Factors affecting hemolysis rates in blood samples drawn from newly placed IV sites in the emergency department. J Emerg Nurs 2005;31:338–445.
65. Lippi G, Salvagno GL, Brocco G, Guidi GC. Preanalytical variability in laboratory testing: influence of the blood drawing technique. Clin Chem Lab Med 2005;43:319–325.
66. Sonntag O. Hemolysis as an interference factor in clinical chemistry. J Clin Chem Clin Biochem 1986;24:127–139.
67. Favaloro EJ, Lippi G. Discard tubes are sometimes necessary when drawing samples for hemostasis. Am J Clin Pathol 2010;134(5):851.
68. Lippi G, Salvagno GL, Montagnana M, Brocco G, Cesare Guidi G. Influence of the needle bore size used for collecting venous blood samples on routine clinical chemistry testing. Clin Chem Lab Med 2006;44:1009–1014.
69. Lippi G, Salvagno GL, Montagnana M, Poli G, Guidi GC. Influence of the needle bore size on platelet count and routine coagulation testing. Blood Coagul Fibrinolysis 2006;17:557–561.
70. Clinical Laboratory Standards Institute. CLSI H4-A6. Procedures and Devices for the Collection of Diagnostic Blood Specimen by Skin Puncture; Approved Standard—Sixth Edition, Vol. 24, No. 21, 2008.
71. Michaelsson M, Sjolin S. Haemolysis in blood samples from newborn infants. Acta Paediatr Scand 1965;54:325–330.
72. Meites S, Lin SS, Thompson C. Studies on the quality of specimens obtained by skin puncture of children. 1. Tendency to haemolysis, and haemoglobin and tissue fluid as contaminants. Clin Chem 1981;27:875–878.
73. Kazmierczak SC, Robertson AF, Briley KP. Comparison of hemolysis in blood samples collected using an automatic incision device and a manual lance. Arch Pediatr Adolesc Med 2002;156:1072–1074.
74. Paes B, Janes M, Vegh P, LaDuca F, Andrew M. A comparative study of heel-stick devices for infant blood collection. Am J Dis Child 1993;147:346–348.
75. Sasaki M. A fundamental concept for development of sample transportation system in clinical laboratory. Rinsho Byori 1993;95(Suppl):S1–S6.
76. Laessig RH, Indriksons AA, Hassemer DJ, Paskey TA, Schwartz TH. Changes in serum chemical values as a result of prolonged contact with the clot. Am J Clin Pathol 1976;66:598–604.
77. Lippi G, Lima-Oliveira G, Coutino Nazer S, Lopes Moreira ML, Macedo Souza RF, Salvagno GL, et al. Suitability of a transport box for blood sample shipment over a long period. Clin Biochem 2011;44:1028–1029.
78. Fernandes CMB, Worster A, Hill S, McCallum C, Eva K. Root cause analysis of delays for laboratory tests on emergency department patients. Can J Emerg Med 2004;6:116–122.
79. Green M. Successful alternatives to alternate site testing. Use of a pneumatic tube system to the central laboratory. Arch Pathol Lab Med 1995;119:943–947.
80. Guss DA, Chan TC, Killeen JP. The impact of a pneumatic tube and computerized physician order entry on laboratory turnaround time. Ann Emerg Med 2008;51:181–185.
81. Keshgegian AA, Bull GE. Evaluation of a soft-handling computerized pneumatic tube specimen delivery system. Effects on analytical results and turnaround time. Am J Clin Pathol 1992;97:535–540.

82. Zaman Z, Demets M. Blood gas analysis: POCT versus central laboratory on samples sent by a pneumatic tube system. Clin Chim Acta 2001;307:101–106.

83. Wallin O, Soderberg J, Grankvist K, Jonsson AP, Hultdin J. Preanalytical effects of pneumatic tube transport on routine haematology, coagulation parameters, platelet function and global coagulation. Clin Chem Lab Med 2008;46:1443–1449.

84. Pragay DA, Edwards L, Toppin M, Palmer RR, Chilcote ME. Evaluation of an improved pneumatic-tube system suitable for transportation of blood specimens. Clin Chem 1974;20:57–60.

85. Pragay DA, Fan P, Brinkley S, Chilcote ME. A computer directed tube system: is effect on specimens. Clin Biochem 1980;13:259–261.

86. Ellis G. An episode of increased hemolysis due to a defective pneumatic air tube delivery system. Clin Biochem 2009;42:1265–1269.

87. Mensel B, Wenzel U, Roser M, Lüdemann J, Nauck M. Considerably reduced centrifugation time without increased hemolysis: evaluation of the new BD Vacutainer SSTTMII Advance. Clin Chem 2007;53:794–795.

88. Ismail A, Shingler W, Seneviratne J, Burrows G. In vitro and in vivo haemolysis and potassium measurement. Br Med J 2005;330:949.

89. Lammers M. Gressner AM. Immunonephelometric quantification of free haemoglobin. J Clin Chem Clin Biochem 1987,25:363–367.

90. Wood WG, Kress M, Meissner D, Hanke R, Reinauer H. The determination of free and protein-bound haemoglobin in plasma using a combination of HPLC and absorption spectrometry. Clin Lab 2001;47:279–288.

91. Lee W, Kim Y, Lim J, Kim M, Lee EJ, Lee A, et al. Rapid, sensitive diagnosis of hemolytic anemia using antihemoglobin antibody in hypotonic solution. Ann Clin Lab Sci 2002; 32:37–43.

92. Blank DW, Kroll MH, Ruddel ME, Elin RJ. Hemoglobin interference from in vivo hemolysis. Clin Chem 1985;31:1566–1569.

93. Thomas L. Haemolysis as influence and interference factor. eJIFCC vol 13 no 4. Available at: http://www.ifcc.org/ejifcc/vol13no4/130401002.htm.

94. Selby C. Interference in immunoassay. Ann Clin Bioch 1999;36:704–721.

95. Sobel AE, Snow SD. The estimation of serum vitamin A with activated glycerol dichlorohydrin. J Biol Chem 1947;171:617–632.

96. Utley MH, Brodovsky ER, Pearson WW. Hemolysis and reagent purity as factors causing erratic results in the estimation of vitamin A and carotene in serum by the Bessey-Lowry method. J Nutr 1958;66:205–215.

97. Hawkins RC. Poor knowledge and faulty thinking regarding hemolysis and potassium elevation. Clin Chem Lab Med 2005;43:216–220.

98. Frank JJ, Bermes EW, Bickel MJ, Watkins BF. Effect of in vitro hemolysis on chemical values for serum. Clin Chem 1978;24:1966–1970.

99. Blank DW, Kroll MH, Ruddel ME, Elin RJ. Hemoglobin interference from in vivo hemolysis. Clin Chem 1985;31:1566–1569.

100. Sonntag O. Haemolysis as an interference factor in clinical chemistry. J Clin Chem Clin Biochem 1986;24:127–139.

101. Randall AG, Garcia-Webb P, Beilby JP. Interference by haemolysis, icterus and lipaemia in assays on the Beckman Synchron CX5 and methods for correction. Ann Clin Biochem 1990;27(Pt 4):345–352.

102. Yücel D, Dalva K. Effect of in vitro hemolysis on 25 common biochemical tests. Clin Chem 1992;38:575–577.

103. Jay DW, Provasek D. Characterization and mathematical correction of hemolysis interference in selected Hitachi 717 assays. Clin Chem 1993;39:1804–1810.

104. Grafmeyer D, Bondon M, Manchon M, Levillain P. The influence of bilirubin, haemolysis and turbidity on 20 analytical tests performed on automatic analysers. Results of an interlaboratory study. Eur J Clin Chem Clin Biochem 1995;33:31–52.
105. Hübl W, Wejbora R, Shafti-Keramat I, Haider A, Hajdusich P, Bayer PM. Enzymatic determination of sodium, potassium, and chloride in abnormal (hemolyzed, icteric, lipemic, paraproteinemic, or uremic) serum samples compared with indirect determination with ion-selective electrodes. Clin Chem 1994;40:1528–1531.
106. Lippi G, Salvagno GL, Montagnana M, Brocco G, Guidi GC. Influence of hemolysis on routine clinical chemistry testing. Clin Chem Lab Med 2006;44:311–316.
107. Steen G, Vermeer HJ, Naus AJ, Goevaerts B, Agricola PT, Schoenmakers CH. Multicenter evaluation of the interference of hemoglobin, bilirubin and lipids on Synchron LX-20 assays. Clin Chem Lab Med 2006;44:413–419.
108. Koseoglu M, Hur A, Atay A, Cuhadar S. Effects of hemolysis interference on routine biochemistry parameters. Biochem Med 2011;21:79–85.
109. Jay D, Provasek D. Characterization and mathematical correction of hemolysis interference in selected Hitachi 717 assays. Clin Chem 1993;39:1804–1810.
110. Kroll MH, Elin RJ. Interference with clinical laboratory analyses. Clin Chem 1994; 40:1996–2005
111. Caraway WT. Chemical and diagnostic specificity of laboratory tests. Am J Clin Pathol 1961;37:445–464.
112. Algeciras-Schimnich A, Cook WJ, Milz TC, Saenger AK, Karon BS. Evaluation of hemoglobin interference in capillary heel-stick samples collected for determination of neonatal bilirubin. Clin Biochem 2007;40:1311–1316.
113. Gobert De Paepe E, Munteanu G, Schischmanoff PO, Porquet D. Haemolysis and turbidity influence on three analysis methods of quantitative determination of total and conjugated bilirubin on ADVIA 1650. Ann Biol Clin (Paris) 2008;66:175–182.
114. Dimeski G, Mollee P, Carter A. Increased lipid concentration is associated with increased hemolysis. Clin Chem 2005;51:2425.
115. Vassault A, Grafmeyer D, Naudin C, Dumont G, Bailly M, Henny J, et al. Protocol for the validation of methods. Ann Biol Clin;1986:44:686–745.
116. Vassault A, Grafmeyer D, de Graeve J, Cohen R, Beaudonnet A, Bienvenu J. Quality specifications and allowable standards for validation of methods used in clinical biochemistry. Ann Biol Clin 1999;57:685–695.
117. Lippi G, Salvagno GL, Ippolito L, Franchini M, Favaloro EJ. Shortened activated partial thromboplastin time: causes and management. Blood Coagul Fibrinolysis 2010;21: 459–463.
118. Favaloro EJ, Lippi G. Laboratory reporting of hemostasis assays: the final post-analytical opportunity to reduce errors of clinical diagnosis in hemostasis? Clin Chem Lab Med 2010;48:309–321.
119. Laga AC, Cheves TA, Sweeney JD. The effect of specimen hemolysis on coagulation test results. Am J Clin Pathol 2006;126748–126755.
120. Adcock DM, Hoefner DM, Kottke-Marchant K, Marlar RA, Szamosi DI, Warunek DJ. Collection, Transport, and Processing of Blood Specimens for Testing Plasma-Based Coagulation Assays and Molecular Hemostasis Assays: Approved Guideline-Fifth Edition. Clinical Laboratory Standards Institute. Wayne, PA: CLSI document H21-A5; 2008.
121. Bauer NB, Eralp O, Moritz A. Effect of hemolysis on canine kaolin-activated thromboelastography values and ADVIA 2120 platelet activation indices. Vet Clin Pathol 2010;39:180–189.
122. Favaloro EJ, Lippi G, Franchini M. Contemporary platelet function testing. Clin Chem Lab Med 2010;48(5):579–598.

123. McGlasson DL, Fritsma GA. Whole blood platelet aggregometry and platelet function testing. Semin Thromb Hemost 2009;35:168–180.

124. Cattaneo M. Light transmission aggregometry and ATP release for the diagnostic assessment of platelet function. Semin Thromb Hemost 2009;35(2):158–167.

125. Favaloro EJ. Clinical Utility of the PFA-100. Semin Thromb Hemost 2008;34:709–733.

126. Gottschall JL, Collins J, Kunicki TJ, Nash R, Aster RH. Effect of hemolysis on apparent values of platelet-associated IgG. Am J Clin Pathol 1987;87:218–222.

127. Kricka LJ. Interferences in immunoassay–still a threat. Clin Chem 2000;46:1037–1038.

128. Wenk RE. Mechanism of interference by hemolysis in immunoassays and requirements for sample quality. Clin Chem 1998;44:2554.

129. Evans MJ, Livesey JH, Ellis MJ, Yandle TG. Effect of anticoagulants and storage temperatures on stability of plasma and serum hormones. Clin Biochem 2001;34:107–112.

130. Snyder JA, Rogers MW, King MS, Phillips JC, Chapman JF, Hammett-Stabler CA. The impact of hemolysis on Ortho-Clinical Diagnostic's ECi and Roche's elecsys immunoassay systems. Clin Chim Acta 2004;348:181–187.

131. Lippi G, Avanzini P, Dipalo M, Aloe R, Cervellin G. Influence of hemolysis on troponin testing: studies on Beckman Coulter UniCel Dxl 800 Accu-TnI and overview of the literature. Clin Chem Lab Med 2011;49: 2097–2100.

132. Lyon ME, Ball CL, Krause RD, Slotsve GA, Lyon AW. Effect of hemolysis on cardiac troponin T determination by the Elecsys 2010 immunoanalyzer. Clin Biochem 2004;37: 698–701.

133. Sodi R, Darn SM, Davison AS, Stott A, Shenkin A. Mechanism of interference by haemolysis in the cardiac troponin T immunoassay. Ann Clin Biochem 2006;43:49–56.

134. Masimasi N, Means RT Jr. Elevated troponin levels associated with hemolysis. Am J Med Sci 2005;330:201–203.

135. Li A, Brattsand G. Stability of serum samples and hemolysis interference on the high sensitivity troponin T assay. Clin Chem Lab Med 2011;49:335–336.

136. Dasgupta A, Wells A, Biddle DA. Negative interference of bilirubin and hemoglobin in the MEIA troponin I assay but not in the MEIA CK-MB assay. J Clin Lab Anal 2001;15:76–80.

137. Hawkins RC. Hemolysis interference in the ortho-clinical diagnostics vitros ECi cTnI assay. Clin Chem 2003;49:1226.

138. Daves M, Salvagno GL, Cemin R, Gelati M, Cervellin G, Guidi GC, et al. Influence of haemolysis on routine laboratory cardiac markers testing. Clin Lab 2012;58:333–336.

139. Florkowski C, Wallace J, Walmsley T, George P. The effect of hemolysis on current troponin assays–a confounding preanalytical variable? Clin Chem 2010;56:1195–1197.

140. Meites S. Reproducibly simulating hemolysis, for evaluating its interference with chemical methods. Clin Chem 1973;19:1319.

141. Bais R. The effect of sample hemolysis on cardiac troponin I and T assays. Clin Chem 2010;56:1357–1359.

142. National Academy of Clinical Biochemistry Laboratory Medicine Practice Guidelines: Use of cardiac troponin and B-type natriuretic peptide or n-terminal pro B-type natriuretic peptide for etiologies other than acute coronary syndromes and heart failure. Clin Chem 2007;53:2086–2096.

143. Jones AM, Honour JW. Unusual results from immunoassays and the role of the clinical endocrinologist. Clin Endocrinol 2006;64:234–244.

144. Steen G, Klerk A, Laan K, Eppens EF. Evaluation of the interference due to haemoglobin, bilirubin and lipids on Immulite 2500 assays: a practical approach. Ann Clin Biochem 2011;48:170–175.

145. Cook PR, Glenn C, Armston A. Effect of hemolysis on insulin determination by the Beckman Coulter Unicell DXI 800 immunoassay analyzer. Clin Biochem 2010;43: 621–622.

146. Kwon HJ, Seo EJ, Min KO. The influence of hemolysis, turbidity and icterus on the measurements of CK-MB, troponin I and myoglobin. Clin Chem Lab Med 2003;41:360–364.

147. Verfaillie CJ, Delanghe JR. Hemolysis correction factor in the measurement of serum neuron-specific enolase. Clin Chem Lab Med 2010;48:891–892.

148. Marangon K, O'Byrne D, Devaraj S, Jialal I. Validation of an immunoassay for measurement of plasma total homocysteine. Am J Clin Pathol 1999;112:757–762.

149. Bossuyt X, Blanckaert N. Evaluation of interferences in rate and fixed-time nephelometric assays of specific serum proteins. Clin Chem 1999;45:62–67.

150. Moalem J, Ruan DT, Farkas RL, Shen WT, Gosnell JE, Miller S, Duh QY, Clark OH, Kebebew E. Hemolysis falsely decreases intraoperative parathyroid hormone levels. Am J Surg 2009;197:222–226.

151. Lippi G, Plebani M. Identification errors in the blood transfusion laboratory: a still relevant issue for patient safety. Transfus Apher Sci 2011;44:231–233.

152. Tanabe P, Kyriacou DN, Garland F. Factors affecting the risk of blood bank specimen hemolysis. Acad Emerg Med 2003;10:897–900.

153. Laga A, Cheves T, Maroto S, Coutts M, Sweeney J. The suitability of hemolyzed specimens for compatibility testing using automated technology. Transfusion 2008;48:1713–1720.

154. Lewis VJ, Thacker WL, Mitchell SH, Baer GM. A new technic for obtaining blood from mice. Lab Anim Sci 1976;26:211–213.

155. Suber RL, Kodell RL. The effect of three phlebotomy techniques on hematological and clinical chemical evaluation in sprague-dawley rats. Vet Clin Pathol 1985;14:23–30.

156. Aasland KE, Skjerve E, Smith AJ. Quality of blood samples from the saphenous vein compared with the tail vein during multiple blood sampling of mice. Lab Anim 2010;44:25–29.

157. Christensen SD, Mikkelsen LF, Fels JJ, Bodvarsdóttir TB, Hansen AK. Quality of plasma sampled by different methods for multiple blood sampling in mice. Lab Anim 2009;43:65–71.

158. Martínez-Subiela S, Cerón JJ. Effects of hemolysis, lipemia, hyperbilirrubinemia, and anticoagulants in canine C-reactive protein, serum amyloid A, and ceruloplasmin assays. Can Vet J 2005;46:625–629.

159. Lucena R, Moreno P, Pérez-Rico A, Ginel PJ. Effects of haemolysis, lipaemia and bilirubinaemia on an enzyme-linked immunosorbent assay for cortisol and free thyroxine in serum samples from dogs. Vet J 1998;156:127–131.

160. Martínez-Subiela S, Tecles F, Montes A, Gutiérrez C, Cerón JJ. Effects of haemolysis, lipaemia, bilirubinaemia and fibrinogen on protein electropherogram of canine samples analysed by capillary zone electrophoresis. Vet J 2002;164:261–268.

161. Glick MR, Ryder KW, Glick SJ, Woods JR. Unreliable visual estimation of the incidence and amount of turbidity, hemolysis, and icterus in serum from hospitalized patients. Clin Chem 1989;35:837–839.

162. Hawkins R. Discrepancy between visual and spectrophotometric assessment of sample haemolysis. Ann Clin Biochem 2002;39:521–522.

163. Simundic AM, Nikolac N, Ivankovic V, Ferenec-Ruzic D, Magdic B, Kvaternik M, et al. Comparison of visual vs. automated detection of lipemic, icteric and hemolyzed specimens: can we rely on a human eye? Clin Chem Lab Med 2009;47:1361–1365.

164. Jeffery J, Sharma A, Ayling RM. Detection of haemolysis and reporting of potassium results in samples from neonates. Ann Clin Biochem 2009;46(Pt 3):222–225.

165. Darby D, Broomhead C. Interference with serum indices measurement, but not chemical analysis, on the Roche Modular by Patent Blue V. Ann Clin Biochem 2008;45:289–292.

166. Vermeer HJ, Steen G, Naus AJ, Goevaerts B, Agricola PT, Schoenmakers CH. Correction of patient results for Beckman Coulter LX-20 assays affected by interference due to hemoglobin, bilirubin or lipids: a practical approach. Clin Chem Lab Med 2007;45:114–119.
167. Sonntag O, Glick MR. Serum-Index und Interferogramm – Ein neuer Weg zur Prüfung und Darstellung von Interferenzen durch Serumchromogene. Lab Med 1989;13:77–81.
168. Lippi G, Mattiuzzi C, Plebani M. Event reporting in laboratory medicine. Is there something we are missing? MLO Med Lab Obs 2009;41:23.
169. Sciacovelli L, Plebani M. The IFCC Working Group on laboratory errors and patient safety. Clin Chim Acta 2009;404:79–85.
170. Söderberg J, Jonsson PA, Wallin O, Grankvist K, Hultdin J. Haemolysis index–an estimate of preanalytical quality in primary health care. Clin Chem Lab Med 2009;47:940–944.
171. Lippi G, Luca Salvagno G, Blanckaert N, Giavarina D, Green S, Kitchen S, et al. Multicenter evaluation of the hemolysis index in automated clinical chemistry systems. Clin Chem Lab Med 2009;47:934–939.
172. Golf SW, Schneider S, Friemann E, Temme H, Roka L. Correction of catalytic activities of aspartate-aminotransferase, lactate-dehydrogenase, acid-phosphatase and potassium concentration in hemolytic plasma by determination of hemoglobin concentration with direct spectrophotometry. J Clin Chem Clin Biochem 1985;23:585.
173. Thomas C, Thomas L. Dyshämoglobine. Thomas L, ed. Labor und Diagnose. Frankfurt, Germany:TH Books; 2005:697–701.
174. Unger J, Filippi G, Patsch W. Measurements of free hemoglobin and hemolysis index: EDTA- or lithium-heparinate plasma? Clin Chem 2007;53:1717–1718.
175. Clinical and Laboratory Standard Institute. New project: Use of Serum Indices (Hemolysis, Icterus, Turbidity) in the Clin Chem Laboratory. Available at: http://www.clsi.org/Content/NavigationMenu/Volunteers/VolunteerOpportunities/CallforNominations/May_09_Chem.htm.
176. Plebani M, Lippi G. Hemolysis index: quality indicator or criterion for sample rejection? Clin Chem Lab Med 2009;47:899–902.
177. Vermeer HJ, Thomassen E, de Jonge N. Automated processing of serum indices used for interference detection by the laboratory information system. Clin Chem 2005;51:244–247.
178. Lovelock JE. The haemolysis of human red blood-cells by freezing and thawing. Biochim Biophys Acta 1953;10:414–426.
179. Lippi G, Musa R, Aloe R, Mercadanti M, Pipitone S. Influence of temperature and period of freezing on the generation of hemolysate and blood cell lysate. Clin Biochem 2011;44:1267–1269.
180. Glick M, Ryder K, Jackson S. Graphical comparisons of interferences in clinical chemistry instrumentation. Clin Chem 1986;32:470–475.
181. Dimeski G. Effects of hemolysis on the Roche ammonia method for Hitachi analyzers. Clin Chem 2004;50:976–977.
182. Milowska K, Gabryelak T, Lypacewicz G, Tymkiewicz R, Nowicki A. Effect of ultrasound on nucleated erythrocytes. Ultrasound Med Biol 2005;31:129–134.
183. National Committee for Clinical Laboratory Standards. Reference and Selected Procedures for the Quantitative Determination of Hemoglobin in Blood; Approved Standard, 3rd ed. NCCLS document H15-A3. Wayne, PA: NCCLS; 2000.
184. Stadie WC. A method for the determination of methemoglobin in whole blood. J Biol Chem 1920;41:237–241.
185. Drabkin DL, Austin JH. Spectrophotometric studies. II. Preparations from washed blood cells; nitric oxide hemoglobin and sulfhemoglobin. J Biol Chem 1935;112:51–65.

186. Miller WG, Erek A, Cunningham TD, Oladipo O, Scott MG, Johnson RE. Commutability limitations influence quality control results with different reagent lots. Clin Chem 2011;57:76–83.

187. Bland JM, Altman DG. Statistical methods for assessing agreement between two methods of clinical measurement. Lancet 1986;1:307–310.

188. Bland M. An Introduction to Medical Statistics. 2. New York: Oxford University Press; 1995.

189. Dewitte K, Fierens C, Stockl D, Thienpont LM. Application of the Bland-Altman plot for interpretation of method-comparison studies: a critical investigation of its practice. Clin Chem 2002;48:799–801.

190. Hanneman SK. Design, analysis, and interpretation of method-comparison studies". AACN Advanced Critical Care 2008;19:223–234.

191. Cava F, Garcia-Lario JV, Hernandez A, Jimenez CV, Minchinela J, Perich C, et al. Current databases on biologic variation: pros, cons and progress. Scand J Clin Lab Invest 1999;59:491–500.

192. Westgard JO, Burnett RW. Precision requirements for cost-effective operation of analytical processes. Clin Chem 1990;36:1629–1632.

193. Lima-Oliveira G, Lippi G, Salvagno GL, Montagnana M, Pitanguiera Manguera CS, et al. New ways to deal with known preanalytical issues: use of transilluminator instead of tourniquet for easing vein access and eliminating stasis on clinical biochemistry. Biochem Med 2011;21:152–159.

194. Lima-Oliveira G, Salvagno GL, Lippi G, Montagnana M, Scartezini M, Picheth G, et al. Elimination of the venous stasis error for routine coagulation testing by transillumination. Clin Chim Acta 2011;412:1482–1484.

195. Frey AM. Drawing blood samples from vascular access devices: evidence-based practice. J Infus Nurs 2003;26:285–293.

196. Owens H, Siparsky G, Bajaj L, Hampers LC. Correction of factitious hyperkalemia in hemolyzed specimens. Am J Emerg Med 2005;23:872–875.

197. Hawkins R. Variability in potassium/hemoglobin ratios for hemolysis correction. Clin Chem 2002;48:796.

198. Brescia V, Tampoia M, Mileti A. Evaluation of factitious hyperkalemia in hemolytic samples: impact of the mean corpuscular hemoglobin concentration. LabMed 2009; 40:224–226.

199. Dimeski G, Clague AE, Hickman PE. Correction and reporting of potassium results in haemolysed samples. Ann Clin Biochem 2005;42:119–123.

200. Shepherd J, Warner MH, Poon P, Kilpatrick ES. Use of haemolysis index to estimate potassium concentration in in-vitro haemolysed serum samples. Clin Chem Lab Med 2006;44:877–879.

201. Carraro P. Potassium report of hemolyzed serum samples. Clin Chem Lab Med 2008;46:425.

202. Lippi G, Banfi G, Buttarello M, Ceriotti F, Daves M, Dolci A, et al. Recommendations for detection and management of unsuitable samples in clinical laboratories. Clin Chem Lab Med 2007;45:728–736.

203. World Health Organization. Use of anticoagulants in diagnostic laboratory: stability of blood, plasma and serum samples. Geneva: WHO, 2002.

204. Salinas M, Flores E, Lugo J, López Garrigós M. Reporting test results in hemolyzed samples from primary care patients. Clin Biochem 2009;42:1204.

205. Scott MG, Kucik DF, Goodnough LT, Monk TG. Blood substitutes: evolution and future applications. Clin Chem 1997;43:1724–1731.

206. Lippi G, Franchini M, Salvagno GL, Guidi GC. Biochemistry, physiology, and complications of blood doping: facts and speculation. Crit Rev Clin Lab Sci 2006;43:349–391.

207. Lippi G, Montagnana M, Franchini M. Ex–vivo red blood cells generation: a step ahead in transfusion medicine? Eur J Intern Med 2011;22:16–19.
208. Lippi G, Franchini M, Banfi G. Red blood cell-mimicking synthetic biomaterial particles: the new frontier of blood doping? Int J Sports Med 2010;31:75–76.
209. Ali AC, Campbell JA. Interference of o-raffinose cross-linked hemoglobin with routine Hitachi 717 assays. Clin Chem 1997;43:1794–1796.
210. Ma Z, Monk TG, Goodnough LT, McClellan A, Gawryl M, Clark T, et al. Effect of hemoglobin- and Perflubron-based oxygen carriers on common clinical laboratory tests. Clin Chem 1997;43:1732–1737.
211. Moreira PL, Lansden CC, Clark TL, Gawryl MS. Effect of Hemopure on the performance of Ektachem and Hitachi clinical analyzers. Clin Chem 1997;43:1790–1791.
212. Ali AC, Mihas CC, Campbell JA. Interferences of o-raffinose cross-linked hemoglobin in three methods for serum creatinine. Clin Chem 1997;43:1738–1743.
213. Kazmierczak SC, Catrou PG, Boudreau D. Simplified interpretative format for assessing test interference: studies with hemoglobin-based oxygen carrier solutions. Clin Chem 1998;44:2347–2352.
214. Chance JJ, Norris EJ, Kroll MH. Mechanism of interference of a polymerized hemoglobin blood substitute in an alkaline phosphatase method. Clin Chem 2000;46:1331–1337.
215. Cameron SJ, Gerhardt G, Engelstad M, Young MA, Norris EJ, Sokoll LJ. Interference in clinical chemistry assays by the hemoglobin-based oxygen carrier, Hemospan. Clin Biochem 2009;42:221–224.
216. Ali AA, Ali GS, Steinke JM, Shepherd AP. Co-oximetry interference by hemoglobin-based blood substitutes. Anesth Analg 2001;92:863–869.
217. Jahr JS, Osgood S, Rothenberg SJ, Li QL, Butch AW, Gunther R, et al. Lactate measurement interference by hemoglobin-based oxygen carriers (Oxyglobin, Hemopure, and Hemolink). Anesth Analg 2005;100:431–436.
218. Osgood SL, Jahr JS, Desai P, Tsukamoto J, Driessen B. Does methemoglobin from oxidized hemoglobin-based oxygen carrier (hemoglobin Glutamer-200) interfere with lactate measurement (YSI 2700 SELECT Biochemistry Analyzer? Oxyglobin) was also assessed by Osgood et al. Anesth Analg. 2005;100:437–439.
219. Moon-Massat PF, Tierney JP, Hock KG, Scott MG. Hitachi Hemolytic Index correlates with HBOC-201 concentrations: impact on suppression of analyte results. Clin Biochem 2008;41:432–435.
220. Kunicka J, Malin M, Zelmanovic D, Katzenberg M, Canfield W, Shapiro P, Mohandas N. Automated quantitation of hemoglobin-based blood substitutes in whole blood samples. Am J Clin Pathol 2001;116:913–919.
221. Dong Q, Stowell CP. Blood substitutes. What they are and how they might be used. Am J Clin Pathol 2002;118(Suppl):S71–S80.
222. Lasne F, Crepin N, Ashenden M, Audran M, de Ceaurriz J. Detection of hemoglobin-based oxygen carriers in human serum for doping analysis: screening by electrophoresis. Clin Chem 2004;50:410–415.
223. Varlet-Marie E, Ashenden M, Lasne F, Sicart MT, Marion B, de Ceaurriz J, Audran M. Detection of hemoglobin-based oxygen carriers in human serum for doping analysis: confirmation by size-exclusion HPLC. Clin Chem 2004;50:723–731.
224. Simitsek PD, Giannikopoulou P, Katsoulas H, Sianos E, Tsoupras G, Spyridaki MH, Georgakopoulos C. Electrophoretic, size-exclusion high-performance liquid chromatography and liquid chromatography-electrospray ionization ion trap mass spectrometric detection of hemoglobin-based oxygen carriers. Anal Chim Acta 2007; 583:223–230.
225. Goebel C, Alma C, Howe C, Kazlauskas R, Trout G. Methodologies for detection of hemoglobin-based oxygen carriers. J Chromatogr Sci 2005;43:39–46.

Index